The Possibilities Mind

Conversations With God, Einstein, and Others

Bernard G. Berenson, Ph.D.
and
Robert R. Carkhuff, Ph.D.

Published by Possibilities Publishing
22 Amherst Road
Amherst, MA 01002
413-253-3488
800-822-2801 (U.S. and Canada)
413-253-3490 (fax)
www.possibilitiespublishing.com

Cover Photo: In this stunning picture of the giant galactic nebula NGC 3603, the crisp resolution of NASA's Hubble Space Telescope captures various stages of the life cycle of stars in one single view.

Credit: Wolfgang Brandner (JPL/IPAC), Eva K. Grebel (Univ. Washington), You-Hua Chu (Univ. Illinois Urbana-Champaign), and NASA

ISBN 0-87425-630-5

Editorial services by Mary George
Production services by Jean Miller
Cover design by Eileen Klockars

To our parents,
who taught us “forever” has an uncompromising social inclination

The Possibilities Mind
Conversations With God, Einstein, and Others

CONTENTS

About the Authors vii
Foreword: *"Window on the Universe"* by David N. Aspy, Ed.D. .. xi
Preface: *"Of Variability and Changeability"* xxi

1. Perspective Is Everything 3
2. From the Leftovers 7
3. $X \neq Y$ Is the Big Bang 15
4. Possibilities 19
5. $X \neq Y$ Is the Source Formula 25
6. Toward God's Formula 29
7. The Possibilities Mind I 33
8. Reinventing Equality 37
9. "Unequality" Is a Hunger 43
10. Historical Social Denial 51
11. The Artificial and Its Terrible Price 57
12. Reinventing Intelligence 63
13. The Asymmetrical Lifeline 71
14. Beyond Chance 77
15. Thank God for Leftovers! 83
16. The Possibilities Mind II 89
17. Perspective Is Born in Relating 97
18. Where There Is a Social Nature, There Are No Voids 103
19. Ideas From the Study of the Possibilities Mind 109
20. "We Do Not Arrive Here Alone" 115
21. Toward Integrative Reinvention 123
22. Freeing Possibilities 133
23. The Masquerade Is Over 141
24. Momentary Functions Toward Meaning 151
25. A Space to Mingle Intellectually 157

References 161
Acknowledgments 163

About the Authors

BERNARD G. BERENSON, PH.D. is Executive Director, Carkhuff Institute of Applied Science and Human Technology. He founded and directed the first Center for Human Resource Development at American International College in 1958. A professor at several major universities, Dr. Berenson is co-author of numerous works on science and psychology, including ***The New Science of Possibilities, The Possibilities Organization,*** and ***The Possibilities Leader.***

ROBERT R. CARKHUFF, PH.D. is Founder and Chairperson of several corporations, including Carkhuff Thinking Systems, Inc. One of the most-cited social scientists of our time, he is also co-author of ***The New Science of Possibilities*** and visionary of other works, including ***The Age of the New Capitalism, Human Possibilities,*** and ***The Possibilities Schools.***

The infinite possibilities universes may reveal their processing complexities only in a scientific community of Possibilities Minds.

7

**In a universe of X = Y,
stasis is everything and
everything is alone.**

The Possibilities Mind I

Einstein believed that the search for a unified source formula is a religious crusade. He wrote: "The most beautiful thing we can experience is the mysterious. It is the source of all art and science."[*] The beauty and power of "the mysterious" live between momentary real limits and infinite possibilities. They are testimony to the generosity of great inequalities.

This is not to say there is no such thing as equality. There is! Equality simply needs to be reinvented, redefined. *Equality is the freedom to participate fully in the merging that occurs from interdependent but unequal asymmetrical processing.* For equality to be fair, participants must have empathy, skills, and the ego-free will to merge in a developmental relationship that creates evolving phenomenal information. That is the prelude to the Possibilities Mind. It is also how the Possibilities Mind contributes to the efficacy of science, for all science is nourished in the womb of rediscovery and, at rare moments, born anew in the reinvention of phenomenal information.

The mysteries of the unequal and asymmetrical entice the Possibilities Mind to process interdependently. These are the adventures of discovery in which phenomena and the possibilities scientist merge to take on their ever-changing, unequal meaning and applications. These mysteries between real limits and infinite possibilities are wondrous because they hold the keys to perpetuating possibilities.

The fragile, tyrannical ego and the probabilities equation are very much like teetering tightrope-walkers. They are dedicated to validating balance, when balance is not an enduring reality.

The eternal conflict is change versus stasis. That conflict will not be resolved by a conventional equation such as $X = Y$, because perfect balance produces nothing at all. The unconventional $X \neq Y$ is at least a poetic and, as such, fertile expression of *what produces everything:* changeability. Here we have the source of new variance, Nature's intolerance for symmetry and singularity, the mysteries upon which feast the Possibilities Mind and all creation.

---

[*] *Living Philosophies* (1949), Vol. 7.

The possibilities $X \neq Y$ validates what Ralph Waldo Emerson referred to as "the privilege of the immeasurable mind"; it gives us the freedom and right to generate phenomenal information. The alternative is to embrace a world such as the one described by T. S. Eliot, in which "the majority of mankind is lazy-minded, incurious, absorbed in vanities, and tepid in emotion, and is therefore incapable of either much doubt or much faith." In this world, Eliot adds, "hell is alone, the other figures in it merely projections. There is nothing to escape to. One is always alone."*

The message of the Possibilities Mind is social, and phenomenal information is the substance of the message. With the Possibilities Mind, one is never alone.

While we have not learned everything about Nature's processes, we have learned more about the sources of Nature's character. *And* we are learning how these sources come together and create the power of a unifying but unequal social process-centricity! In so doing, we are being introduced to Nature's intellect, mind, and perspective.

Again, this is social, even conversational. We find we are not shaping what we are writing and thinking. We are writing and thinking what the voice of Nature tells us to write and think. These conversations require two preconditions:

1. An architecture so tidy it can be driven by Nature's social changeability, and thus be generative and interdependent in its deductive models.
2. A bracketing of the ego so that we can interdependently relate to the emerging phenomenal substance and merge with it.

Together these preconditions help to liberate us from the apparent conflicts between the architecture of science and the seeming untidiness of Nature. While science follows its mission to make things tidy, the dynamics of Nature are never constantly tidy. Nor are they ever equal!

* From Eliot's introduction to Pascal's *Pensées.*

8

Phenomena may merge to form a unity, but the merging is never equal.

Reinventing Equality

The architecture of the new science of possibilities is tidy enough to be interdependent with Nature's changeability, that is, social process-centricity. It need never be absolutely tidy! *A possibilities perspective is both tidy and untidy all at once!* The tidy and untidy absolutely require each other to perpetuate some degree of asymmetry as they accompany each other. The tidy and untidy are never equal; the momentary real limits and infinite possibilities are never equal. Social process-centricity itself is never equal for very long, if ever.

Notice that here, unlike probabilities thinkers, we do not attempt to fit science and Nature into a comfortable place within our perspective, neglecting the notion that Nature may have its own perspective. We acknowledge and seek to merge with Nature's perspective, which is a possibilities perspective.

The power of this approach is reflected in the following points:

- The architecture of the new science is socially process-centric, a core quality that makes a possibilities perspective operational; thus, the new science may be the true and faithful translator of Nature's social process-centricity.

- Possibilities' architectural interdependency and merging with Nature provides *a sociogenetic place* for Nature to share its possibilities perspective through its voice of changeability.

- The new science's architecture and Nature's possibilities perspective are both born and nourished in the womb of unequal social forces. The new science thus realizes the enormous significance of such genesis: *it grants Nature its powers to create the evolving cosmic melting pot out of all the connected and connecting universes that are and those yet to be.*

- The possibilities architecture images the dynamic connections among and between the relating, empowering, and freeing powers of changeability: changeability → unequal social process-centricity. This reveals the sense and significance of Nature's growing openness to its own tolerance *and* intolerance for short-lived polarities: relativity and chaos, symmetry

> and asymmetry, singularity and diversity, as well as relative gravity in all the complex and seemingly contradictory influences that flow from its own fertile irregular and social character.
>
> Indeed, in this context, possibilities science realizes that it is this simultaneous but unequal tolerance *and* intolerance for polarities, the apparent and real, that allows opposing phenomena to operate at the same instance in the same context. Functionally and phenomenally, they are companions and, in that sense, short-lived polarities.

As all the above makes clear, even ambitious revivals of learning cannot impact nor fully employ logic, reason, and empathy when science is not fully focused on *relationships and their possibilities.* How phenomena relate to and with everything else, and how the resulting changes generate infinitely more truths than the study of the phenomena by themselves, is imaged in the new science of possibilities.

All of the history of science is about coming to this revelation. Without it, *all* Nature's secrets would remain hidden. There are no natural truths without it.

The new science is about how universal social forces live and act in and between *connections;* connections between the absolutely finite and absolutely infinite; between absolute simplicity and absolute complexity; between the present, past, and future; between the appearance of linearity and the reality of the changing cycles of the winds of interdependency; between the limits of probability and the liberty of possibilities.

The challenge for the new scientist-biographers is to discover what it is like to be inside these mutating connections—inside the mind of changeability, inside the mind of Nature. They will discover that Nature and the new science are chameleon-like: one determines the changing processes of time, space, relationships, connections, mergings, and structures as functions of their social character; the other emerges out of relating with these processes by merging fully with them and their changing social relationships.

The social character of Nature is the *"determiner."* The new science is the ego-free relating *"mergerer."*

The repetitive and too often immature visions of history and political-economic-social perspective are accounted for by a blindness to the insight that free phenomena are governed by the power of an unequal social process-centricity. This is how Nature thinks! Probabilities science stops short of understanding this.

Social process-centricity is Nature's mind, arms, legs, method, blood, and spirit. It is the dynamic of relative gravity and thus the cosmic fabric that makes it possible for all universes to be connected. Expressed by relative gravity, unequal social process-centricity forms the *"connective changeability bonding tissue"* (CCBT) between universes and dimensions of time. CCBT creates the pathways to and through the thresholds of possibilities. The interdependent changeability models of the new science allow the selection of the thresholds, instant to passing instant, establishing links between cosmic "growth" requirements and cosmic changeability.

CCBT is the passageway between universes and time travel.

Even as universes converge, disperse, and merge, cycle after cycle, they are sociogenetically connected. They reflect one another as they change both in their own ways and in their unequal relationships. In this way, they are *"one"* even as they exercise their unique developing, diverse time-space dimensions. *They are engaged as they reflect upon one another. This socially reflective and merging relationship, sociogenetically determined, creates new changeability dimensions. This is cosmic magic!*

The new dimensions are variable expressions of the basic *asymmetrical social character of Nature.* Although Nature tolerates symmetrical and singular cosmic moments, as well as the asymmetrical, there is slightly more *intolerance* for symmetrical phenomena and singularity. Without this miniscule difference, there would be no universes.

This difference is the source of all matter—of everything!

Even matter and antimatter mirror one another. They are connected. There cannot be one without the other. Matter and antimatter enjoy the most intimate relationship in all the universes. However, they are not perfectly equal!

When they meet, both cease to be. What remains is an instant of the purest, most powerful energy—the most generative energy plus a very small number of protons. All universes thus owe their very existence to this slightly unbalanced, asymmetrical relationship between matter and antimatter. *This creative, productive, social process-centric and unbalanced interdependence is the genesis of asymmetrical universes.*

Social process-centricity, asymmetrical universes, and the new science are born again and again at these same *"most generative energy instants."* They create changeability reality from the unequal tolerance for the symmetrical and singular and the resulting asymmetrical and diverse.

All things share with all creation not only contrasts but also connection: the forever asymmetrical interdependence. This is where change comes from! Again, this is where everything comes from! What is to come is given life by this same small imbalance, yielding and shaping asymmetrical social Nature.

These evolving, emerging universes enjoy a dynamic connection pulsating with possibilities within and between possibilities—new dimensions, different places all at the same moment. *And so it is for all of us!*

The basic processing character of social process-centricity—and thus changeability, relative gravity, CCBT, generative energy instants, and asymmetrical interdependence—is set in motion by the common *unequal sociogenic* DNA of Nature.

Our mentor and friend, Jim Drasgow, states the basic implication of this clearly: *"All humans are born unequal so we can help one another."*

9

From chaos to order,
From order to chaos,
From blindness to clarity,
From confusion to certainty,
From tears of disorientation to the
unbridled joy in the journey.

"Unequality" Is a Hunger

Inequality, or "unequality," is the precondition and the essence of social process-centricity. The social inclination can only be expressed in asymmetrical interdependence. Asymmetrical interdependence is the boiling caldron of process-centricity. Changing, non-parametric universes require the unequal, the asymmetrical. All universal relationships within and between phenomena are unequal. They are the very fabric of the connective changeability bonding tissue, or CCBT, allowing time and space transitions made up of interdependent, unequal processing fibers.

The eternally developing, maturing cosmos introduces so many complexities and so many liberties that relating to it is like brawling with the most powerful headwinds ever. Yet, the struggle is strangely welcoming, for we come upon new insights. Recognizing that asymmetrical interdependence and social process-centricity are more than companions is one of the great prizes. They are the nature of the mergings and unequal unities that account for continuously evolving phenomena, the result of the socio-cosmogenic legacy of all creation. They are enormously freeing.

Whereas asymmetrical interdependence is the fundamental form and structure, social process-centricity is how Nature functions. If the processing were equal, there would be no leftovers and, therefore, nothing. Unequal social process-centricity is the enduring, evolving groundwork for changeability.

Now we understand why a natural physics cannot rest upon X = Y. In Nature, in human affairs, the fluctuating interdependence between phenomena is rarely balanced. Consequently, we find it more and more comfortable to record what we see when Nature shares its vision of an asymmetrical and even non-parametric creation. It is ego-free and thus censor-free: we hear the sound, rhythms, and unfamiliar melodies of meaning emanating from the social process-centric alliances.

In these privileged moments, we are not bound by fixed parameters. We are introduced and reintroduced to constantly reinvented relationships and members of the new alliances. We change rapidly. We are reinvented. All entities are linked by their shared CCBT pathways. Never tranquil, never conceptual, never in equilibrium, the affiliations mutate in an increasingly empowering and freeing socio-cosmogenic process. Bound together in this evolving manner, we become the quintessential community. Each processing member is now able to keep probability in proper perspective: probability is the lure for the losing gambler who offers a single resource not only finite but controlled by only a few. Each member now moves on to fulfill his, her, or its unique cycles of rebirth.

We are no longer combatants in a battle with the headwinds. We are biographers. What we see and what we hear is multisensual, phenomenal cosmic magic, marvelous with evolutionary and revolutionary truth.

From the unforgiving shadow of empty space to the innocence of the first color of morning light, we race from the fuming cyclone to the welcomed cooling breezes of midsummer. We ride the waves of gloriously painted early evening skies to the innumerable multitudes populating the plentifully seeded space beyond sight. We owe this, the *companionable* trait of personality (however defiant, however garrulous), to social process-centricity: *everything is social,* including states of physics and states of mind, heart, and soul.

In this great contentiously gathering party, humankind is the singular temporary tenant who is struggling; but as yet it is too insignificant to spoil that party.

Humans, Nature's biographers, are learning that even solitude is another season for relating!

Everything is a hunger for relating!
This, too, is ***cosmic*** *magic.*

Captured in the season of solitude, memory hungers for relating: isolation falsely satisfies appetite. Curiosity becomes hunger itself, while knowledge becomes evidence of the hunger. In this forgiving season of humankind's solitude, change becomes the hunger for what is new and interdependence becomes the hunger for something more.

Season after season, processing is the hunger for life and centricity the hunger for the merging that intensifies. And season after social season, technology is the hunger for being useful and science the hunger for insights about relationships. All the while, the new science is the hunger for possibilities and social cosmogenic freedom.

Tragically, catastrophically, the rewards of these infinite degrees of freedom are lost in the linear and the probable. Sadly, the fear of freedom's changing possibilities may well be the most telling lesson we can learn from human history and from our collective psychology. Interchangeable responses are nowhere to be heard!

These freedoms, these prospective liberties, provide no stability. If they did, they would not be freedoms; at best, they would be empty promises. As part of the genesis of possibilities, they stimulate and respond to the healthy appetite for expanding perspective about Nature and about humankind. Expanding perspective may be the only experience that will rescue all of us from the confines of probability-based finite resources and, thus, the trappings of a conditioned, incentive-driven level of motivation.

Such motivation sustains conditioned mentalities that are bound and bonded to a two-response, *"Go/No Go"* repertoire. A two-response repertoire is all that is left to transform perspective into empowering, socially connected wisdom. It will not do. It is hopeless! It is an inhibitor of potential, of all the constructive and enduring insights and initiatives that would develop from accelerating change. There is nothing positive about such constraining options.

Nietzsche wrote, "When people are devoid of purpose, the void becomes the purpose." We might add that the same happens when people are devoid of legitimate purpose. Stability—an impossibility—is more than freedom's enemy: as the purpose for anything, stability is a void. It is no secret that defenders of this void commit every imaginable cruelty in its name once they think they can be in control. There are no emancipating options, no possibilities.

The problem:

> *"There are things killing me, and they are all intimately social. The more I try to relate to them, the more rapidly I feel I am dying. It should not be this way, but it is, for me. The intimately social should be free, but it is not for me. So, I know, I will die much sooner than much later."*

The solution:

> *"The things that nourish vitality and life are also intimately social. It should be this way, and it is! The intimately social is the source of change for the universe and for you. Change is the inheritance of freedom, and freedom the legacy of change. The intimately social is interdependent and unequal. Learn this, understand it, and you will not die so soon. Then you may do what is required to live."*

As our friend Jack Kelly said, "We are at war to free ourselves of the restrictive burdens of our old responses. Our weapons are our thinking skills. Our destiny is in our minds."

Our old conditioned two-response repertoire, linear and probable, continues as the rule. It leaves no opportunity to explore inclusively, to understand interchangeably with dynamic phenomena, and no opportunity to engage Nature and God in conversation to request what we need to know in order to partake of socio-cosmogenic freedom. The question can only be presented without ego and free of shame from a conditioned two-response mentality.

Question:

> *"What do we need to know to be part of socio-cosmogenic freedom?"*

God's response:

> *"As it is with Nature, so it is with all the universes and the human brain: social process-centricity is self-organizing."*

Thus the life and mind and personality and perspective that Nature shares so diversely is beyond traditional scientific specialties. Nature, the universes, and the brain are far too dramatic, and their connections, though ever changing, are central to their self-organizing propensity.

While the voice of Nature is the ultimate instrument for expressing complex and dramatic harmony, its message is the social fabric for the composition of the harmony. The new scientist is simply the explorer-recorder. The relating, empowering, and freeing counsel is the *"new science symphony"* within which the harmonious voice of Nature is nested, developed, orchestrated, and performed. The chaotic *and* the determined universes are welcomed as one soaring voice. The quarantined specialties are now able to relate and fully share their power; we can see that the chaotic and the determined occur simultaneously, overlapping but never merging. Now all cycles of singularity/symmetry/asymmetry go on and on, contributing their changeability statements to the cosmic magic of social process-centricity.

The increasing expansion of the universe (or universes) may well be due to movement toward colliding then merging with another universal entity, a movement like that of arms opening to embrace. Universal entities in their social process-centricity relate, empower, and free each other to be faithful to their own and collective sociocosmogenic freedoms. This will continue for as long as the chaotic and deterministic occur simultaneously with just enough opportunity for matter and antimatter to meet and parent leftovers to add to the asymmetry that creates the changeability, interdependent (but unequal) processing possibilities universes.

The infinite complexity of the universes depends upon leftovers. All things in, on, and around such phenomenal universes are made from leftovers, whatever their expression; even rare human genius.

Nature's unequal tolerance for symmetry and asymmetry is the womb for leftovers. Even gravity is simultaneously "relative" and "chaotic" in its social process-centricity, contributing to the asymmetry that allows sustaining unequal relationships among and between all things.

Thus, whether moving or not, the phenomenal entities of cycles are changing statements of Nature's sociogenic composition.

10

The brain is far more complex, far more dramatic, than the computer.

Historical Social Denial

By insisting upon fitting Nature, even God, into a small and specialized stability-oriented perspective, the human intellect became isolated: less free, less empowered, and less able to relate fully to and with phenomena of any kind. This served the quest for, and defense of, control and stability rooted in linear conditioning. Together, non-relating specialties and risk-laden probability formed the foundation for all major institutions, including science, thus relegating humankind to the narrow linear path of a minor, ineffectual player among enduring social process-centric forces.

A sign of just how narrow that path has become is the brain-computer analogy. Here, the self-organizing power of the brain is neutralized in support of, again, a specialized, stability-oriented conditioned perspective. Certainly, God would not approve of this analogy. In the first place, we humans do what we do because, unlike computers, we are not bound by fixed parameters. Fixed parameters only feign the interchangeability and instant dramatic possibilities of the brain. The human intellect, in its most growthful free state, is non-parametric. That is only one source of its great power to organize, reorganize, and thus reinvent.

When we first asked God, "What do we need to know?" the answer made no reference to the computer. Rather, God told us: *"As it is with Nature, so it is with all the universes, the cosmos, and the human brain. Social process-centricity is self-organizing."*

The neuron, an impulse-conducting cell, does have an "all or none," binary-response pattern. However, unlike the computer, the neuron processes *interdependently* in a great number of relationships with numerous changeable variables in its fluid environment: nutrients, temperature changes, fluctuations of physics, hereditary legacies, degrees of light, glandular activity, as well as electric impulses—to name but a few. Although computers can be programmed to estimate variability, the instant-by-instant changeability values of neurons are beyond the probabilities-based estimates of these values and momentarily fixed parameters.

Also, we can *grow* neurons. The computer cannot grow neurons, for it does not have any! This organic dimension of the brain includes the brain's self-organizing capability to produce countless "programs" by itself. The computer cannot do this as well because it has no immediate access to *the known and unknown changeability of socially self-organizing phenomena.*

Although the brain-computer analogy is sometimes employed to illustrate the power of the brain, the brain is far more powerful than the analogy suggests. Its dramatic possibilities also make it far more interesting, even wondrous; by comparison, the computer is a very boring thing indeed. Because the brain is organic, analogies between organic phenomena are more valid and more demonstrably powerful.

The new science of possibilities is also organic. It draws its power, as does Nature, from the convergence that connects the expanding, narrowing cycles of change that characterize interdependent social processing relationships. The power is thus neither stable nor controlled; it is gained and spent interdependently by means of social process-centricity. Consequently, the new science cannot be limited and bound by any conditioned, ego-centered, two-response-repertoire culture. The power of convergence brought about by social process-centricity—in all its phases, operationalized by its models—reflects *a cosmic culture of freedom.* Conditioned linear mentalities do nothing but arrest development and reject such freedom as well as the accelerated change that freedom generates.

In the very first moment that we contribute to empowering changeability with the architecture of the new science—the very first moment of a clear, simple, and true human legacy—we replace foolhardy efforts to control phenomena based upon *partial variance accountability.* The free changeability resulting from the movement in interdependence is accompanied by, *and* accomplished with, totally merging and synchronized diverging sources of unequal variance. Such countless, diverse merging and diverging events are *required* by *enduring* interdependent but *unequal* social processing relationships. Remember, if all things were *equal* (in the probabilities sense of the word), nothing would change and everything would return to the void: *equality now means sharing in social processing interdependently.* Accordingly, all phenomena are empowered with their own universal freedom so generously bequeathed by the changeability movement of unequal interdependence. The new scientist is endowed with this social cosmogenic freedom. The architecture of the new science images what may be done with it.

Although man may well be a "singular creature" as described by Jacob Bronowski, man is another carrier of the sociogenic and cosmogenic. Perhaps this is why Bronowski goes on to explain, "Man is the ideal explorer of Nature because evolution has not fit man to any specific environment," at least on this planet and perhaps any planet.*

Tapping their social cosmogenic freedom, the new scientists may truly understand that great alterations in how we see our universes are brought about by elevating the descriptions of interdependent relationships so that those relationships are more inclusive, more unifying: they may apply this understanding to the imaging of DNA, the theory of relativity, field theory, evolution, and the merging of mathematics, philosophy, and experimentation. Introduction to the near-universal application of the new science architecture yields new momentum for changeability and its endless transfigurations *because it makes donations to their freedom!*

A larger role, a universal role, for humankind has no authenticity until the new science, in all its *intentional-ideational* vigor, expands the perspective, pursuing the implications of the changeability that emerges in and from unequal social process-centric interdependence.

---

* Bronowski, J. *The Ascent of Man.* Boston/Toronto: Little, Brown and Company, 1973.

11

There may be no natural laws; there are only the temporary, social process-centric practices of Nature.

The Artificial and Its Terrible Price

Probabilities-based history, or history as we have typically known it, is not fully human. Too much of it is artificial. Its repetitive cycles are so periodically brief and common that they are out of step and out of time and outside too many possibilities. Such history is a chronology of the abandonment of freedom for the false promise of stability. Its legacy is out of sorts with universes that have a dynamic social changeability identity.

History is so out of kind, so unnatural, that the pursuit of control and stability has left but a tenuous place for the development of dynamics, and only minor dynamics at that. Long-term positively directed dynamics are too frequently assaulted or, worse, denied and ignored. The consequences haunt us. Apparentness—what is readily seen, what appears to be—is defended in isolated institutions, and often with less-than-honest intentions. *Probabilities science is a relatively isolated institution.*

Cut off from social possibilities, probability designs engender a psychology that sustains the artificial as reality. So nothing—not economics, government, education, science—is real, leaving only the autonomous commercial.

In the world of the artificial, apparentness allows even mathematics to become a game employing symbolic language to fabricate increasing numbers of contrived dimensions. Only a few people on the margin survive to introduce fundamental insights and unifying perspectives. Too often, under the spell of apparentness, they become historical curiosities. What else can social isolation do when confronted with a *glimpse* of cosmic magic?

The apparentness left behind by the artificial is not real simply because it is not free. It is not free because it is controlled. It is not empowered to be free because it is not part of the movement in interdependent processing. The artificial is not part of that movement because it is not constituted to relate fully with anything: *the artificial is not social!*

Apparentness depends upon probabilities. It cannot socially unify. Its utility continues the deficit gamble. Even in the artificial universe of a less-than-human history, the apparent occupies a very fragile corner set upon the risky gambles of probability. Its captured, restricted process is no process at all: it cannot be of generative use to itself or any phenomena with a social character. It is simply designed by conditioned linear mentalities to defend its fixed, independent token-view and influence. By putting probability within the context of the new science, we put apparentness to rest.

Over time and space, the driving social-centric force of all natural phenomena is *to be free in order to be of use,* that is, to contribute to a unifying, changing, and *diverse* journey toward singularity, the fleeting precondition for yielding the asymmetry that is the birthplace for new unequal changeability variance. Phenomena must be free to fulfill their destinies.

Contributing to the new free variance of cosmogenic changeability is the greatest good. Such contributions mature into elevated phenomenal expressions affirming social process-centricity as unequal interdependence. *To be generative, interdependence must be unequal—asymmetrical.* In contrast, social isolation moves toward the void of stability and can only encourage, from all other processes, chronic neglect. Such neglect gives apparent life to unnatural marriages between social-economic-political controls and technologies designed to transform potential information capital back into data that was suspect in the first place.

The new science represents an appetite for unifying insights about the nature of expanding unequal relationships. Again, over time and space, there are no equal relationships.

Unequal relationships are the dynamic center of the social process-centricity that, in turn, is the dynamic center of universal (changeability) principles. These universal principles reinvent the universes *and they reinvent the mind.*

The new science of possibilities provides method, idea, theory, perspective, and applications architecture for the evolving sources of universal development.

The new science unifies all sciences.

The new science renders all science applied science.

The new science generates its own higher levels of interdependent social process-centricity.

The new science's interchangeable responses to Nature positions the science in such a way that it can anticipate Nature's evolving expressions and contribute to and even accelerate their evolution.

12

If contributions to social process-centricity were equal in quality and/or quantity *and* provided fixed parameters, the universes and all their cosmic comrades would not, could not, be connected relatively and/or chaotically. They would not exist at all.

Reinventing Intelligence

The new science of possibilities is a reinvention of intelligence. It is made fertile with its possibilities language. Its maturing becomes a kindred community-processing intelligence. It is the only intelligence that fuels profound unifying perspective. It is perpetually generative and thus changes how the human genius processes elevated ideas and technologies and engineers their interventions.

Such intelligence begins its germination in the efforts to formulate a possibilities architecture of interchangeable responses to all phenomena (*the momentary equals* and *the cosmic unequals*) and to Nature itself. These responses, when accurately interchangeable, compose a unifying and universal language. It is a language with which we can all process. This language vibrates with the diverse cadence of mutating unions and with the vitality of phenomena reinventing the implications of elevated relationships.

Reinventing the implications of elevated, inclusive relationships is the function of the Possibilities Mind. While formulating perspective, the Possibilities Mind processes the new science into responses to the following:

- What, if anything, is uniquely human, and what is the character of its interdependence with what is universal?
- What, then, are the expanding parameters of substantive human and cosmic parameters? Might they be non-parametric?
- What are the sociogenic nesting and rotating asymmetrical dynamics of an evolving possibilities vision of a unified scientific cosmology?
- What drives and sustains the evolution of the social character of the changing universal companions?
- How and why do universal entities process within and between inputs into perpetually generative, diverging and converging functions?
- What is the changing structure and source of the processing vitality of the symbiotic relationships between time, diversity, functions, new variance, and perspective?

- How and why does detailing the primal force of changing interdependent processing facilitate tapping the social character of Nature to formulate renewing and unifying possibilities?

Along the way, the Possibilities Mind offers processed operations for:

- The basic mechanisms accounting for awareness and processing;
- The social insights for unlocking the mysteries of discovery, and the generative expanding of unifying social relationships;
- Paradigms to organize, analyze, align, represent, and act upon change itself to advance all processing substance;
- Changing the criteria for predicting the potential and expected efficacy of new models—social, political, commercial, technical, scientific;
- Welcoming and employing a growing understanding of freedom and its reliance upon empowering and relating;
- Making the interdependence between and among relating, empowering, and freeing commonplace in all processing and initiative;
- Pursuing the implications for humankind when it joins in the natural freedom of the Universe and altered assumptions;
- Transforming irrational independence into rational, natural interdependence;
- Moving beyond mindless conditioning toward elevated and elevating levels of processing;
- Welcoming and exercising burgeoning degrees of freedom;
- Allowing scientists to begin discovery with the same database and the same interdependent summary of that changing data, even when data represents diverse functions, conditions, and changeability standards;
- Redefining and renewing the potency of the relating, empathic individual mind;

- Welcoming change and welcoming the empowerment to move beyond the fear of freedom for all scientists and their civil brothers and sisters.

This is how possibilities scientists record the ripples of unequal social relationships, human and cosmic. The universal, interchangeable syntax of this processing is the pavement for the pathways to our echoing levels of understanding and applications. It is a continuously generous generator. While it frees our interdependent processing, it frees new requirements, revealing altered values. It, too, is unfinished.

Everything is happening all the time, at the same time. We migrate among the events we report until reinvention becomes a working principle. God built reinvention into interdependent social processing as an exhibition of social cosmogenic freedom. Our reinvented values are thus defined by input from social processing rather than by fixed values attached to apparent repetitive cycles attached to fixed parameters.

Like all phenomena in the sociogenic cosmos, the architecture of the new science—*this universal community language*—is self-organizing in this processing context and so reinvents itself. It offers transferable applications for all the specialties of science: physical, social, even economic. With this shared language we can continuously reinvent the marketplace of scientific perspectives to offer the first theory of science. This will help render social-scientific "isms" mere curiosities. Then, interdependent-processing scientists may begin to escape the limiting, conditioned, private messages they hear coming from the voices in their heads, and at least hear and understand the compounding, unifying messages from all the social process-centric universes.

A socially isolated intellect accounts for the inability to apply interdependent process-centricity and for the tunnel vision and characteristic cycles of history.

The new scientists do not need to become "like" Nature. They will move inexorably toward the discovery that they *are* Nature as are all phenomena. They will journey through time and between universes, minds, and spirits faithful to socio-cosmogenic freedom, and, by self-organizing, they will embrace their interdependent minds and souls not monolithically but socially. All the while, these new scientists will contribute to the tipped scale of natural phenomena. They will recognize this tipped scale as Nature's compass pointing toward cosmic solutions.

Within this context, the architecture of the new science will be the universal community language that will serve to unify, as H. G. Wells recommended, three great ideas: science, a universal righteousness, and a human commonwealth where the "minds of rare and exceptional persons and peoples in which they first originated signify a clear vision of man in relation to the things about him." These conditions will go on to give "new colour, then a new spirit, and then a new direction to human affairs."* Universal affairs, too, we may add.** The wonder of the new science is the multitude of other new sciences it generates, thus producing new color, spirit, and direction—possibilities—and new intelligence.

The functional interrelationships and ultimate merging of all systems, models, skills, and supportive knowledge of all specialties ensure that no specialty is destined to become a depressor variable creating a race toward total isolation.

The new science of possibilities and its operational architecture thus offers a socio-cosmogenic, process-centric unifying intelligence:

- This unity is *not*
  - stable,
  - singular,
  - symmetrical,
  - independent.

---

* Wells, H. G. *The Outline of History: Being a Plain History of Life and Mankind, Vol. 1.* Edition revised by Raymond Postgate. New York: Doubleday, 1961.

** We may note that this recommendation from Wells, dating to the early twentieth century, was stated without the great support of a universal community language that images the complex harmonies of cosmic magic.

- This unity *is*
  - asymmetrical,
  - interdependent,
  - chaotic,
  - relative.

  It is cosmogenically free.
  It is self-organizing changeability.
  It is non-parametric.
  It is—as is all social processing—all these dynamics.
  It is all these dynamics all at once, at the same instant.

13

The wonder of the new science is the multitude of science it produces from leftovers.

The Asymmetrical Lifeline

Nature's insistence upon interdependent processing is Nature's lifeline. It is a fundamental message that humankind's social intelligence duplicates Nature. It constitutes the core of genius. To insist otherwise is the essential folly of the assumption of independence, and makes inevitable the finality of isolation. Independence is the fatuity of whatever is ego-driven.

With Nature's maturing intelligence, great insights are born in the rhythms of relating and changing phenomena.

During the Middle Ages, theology was the royalty of intellectual activity. Today, financial profit-seeking is heir to the throne. Limited and thus limiting, private and asocial inductive excursions are its exclusive process. The results are certain to be short-lived whims of chance. The price to break these bonds is empowering and freeing intellectual commerce. Otherwise, we are left with ego-defended concepts in conceptualized marketplaces. Probabilities science falls in line.

For all its contributions, probabilities science does not describe the character of Nature and our world as a whole. It focuses upon small slices, with only occasional links to other slices. Each scientific department selects its private slice, often randomly. Each department becomes a compartment that is increasingly self-conscious, thus inhibiting coherence. The inhibition of developmental levels of coherence continues to serve the controlling royalty very well.

The new science of possibilities demands empathy-driven coherence that penetrates the maturation of reciprocity. Empathic relating is the essential dynamic ingredient for the generative thinking that unravels and unifies the mysteries of interdependent relationships. It forms the fabric of social process-centricity and the production of interdependent, developmental, operational processing models. These operationally detailed processing models reinvent themselves in harmony with Nature's changeability.

When power-seeking and ego-driven excursions offer stability in return for the surrender of empowerment and freedom, they offer equations—apparentness—and so promote, and thrive on, the artificiality of equality. These excursions yield desire without curiosity and hope. The absence of curiosity and hope is the vacuum left behind by the restricted variance of probability. There can be no generative perspective.

Ralph Waldo Emerson, in his essay "Spiritual Laws," wrote: "There is a soul at the centre of Nature and over the will of every man, so that none of us can wrong the universe." We may add that at least none of us can wrong it without paying a God-awful price. In "Circles," Emerson went on: "Beware when the great God lets loose a thinker on this planet. Then all things are at risk." We may add that even equality is among those things.

The new science questions $X = Y$. In the larger context of interdependent processing, the new science replaces $=$ with $\neq$, $>$, or $<$. Except for brief moments of singularity, there is rarely any application for equality ($=$). In the postulates based upon social process-centricity (and by extension, interdependent processing), there are no "equals." The architecture of the new science is developmentally imaged as interdependent-processing paradigms. The new science replaces $=$ with $\leftrightarrow$ and $\rightarrow$, representing interdependency between and among things that may be more or less, but never equal. Again, equality in possibilities terms is simply full participation in unequal interdependent processing.

Interdependence *requires* unequal participants. Interdependence between equals is a contradiction in terms, for the processing inherent in interdependence is negated by the stasis of equality and its link to the symmetrical. Nature's changeability does not tolerate equals, nor does singularity tolerate equals for very long. Developmental, interdependent processing requires asymmetry.

Thus:

- Where there is no asymmetry, there is no interdependence.
- Where there is no interdependence, there is no processing.
- Where there is no processing, there is no change.
- Where there is no change, there is no vitality, no life.
- Where there is no vitality, no life, there is no intellect.

- Where there is no life and intellect, there is only every manner of poverty, of the mind, soul, and heart, for Nature, humans, and science.

The inequality ($X \neq Y$; $X > Y$; $X < Y$) required for interdependent processing and hence social process-centricity, so cardinal to the new science, does not connote superiority of one or other participating, relating phenomena. Inequality is the variable, fundamental, prerequisite condition for all levels of processing: individual, interpersonal, interdependent social process-centricity.

Interdependent processing is the engine of change. Social process-centricity is the mutating and maturing, unifying activity. Processing is inexorably natural. Change is natural. *Change is the performance of a processing Nature.* It is a performance of elevated levels of phenomenal maturity. So it is with intelligence when ego-free enough to duplicate Nature!

The technical architectural details of the new science of possibilities are the new proofs for ($X \neq Y$; $X > Y$; $X < Y$) interdependent processing. This interdependent processing is the changeability foundation for social process-centricity. Unequal interdependent processing (↔) is the essential and primary unifying ingredient for the definitions of everything in God's connected and connecting universes. It may well be the singular unifying ingredient for all the perspectives that have escaped or are clamoring to escape our conditioned minds in order to join together and be free.

When the new scientists are free to join humankind, they will be agents for all the moments to come that are rife with fresh but venerable processing responses, responses to be made new again and again with elevated relationships, as it is with Nature. It will then be possible for all people to become heroic sources of a heightened empathic, social cosmic vision. People will discover this vision by welcoming Nature's reinvention. All isolating neuroses will be vanquished. Then humanity may rise to the challenge of learning how to live and grow with the inclusive, processing, empathic and selfless mind.

Reinvention is thus a privileged changing gift brought to us by process-centric time. This gift configures the image-model for a human possibilities culture. Such a culture, free of isolated, privatized, pathological transgressions, will evolve time and time again in harmony with Nature's maturing cycles and possibilities. This is *a social-physical revolution.* The composition of this revolution is new variance. The reality of the revolution is demonstrated in the maturing cycles of changeability. Their enduring truths are found in the clarity of a constant evolution toward an increasingly unifying processing perspective. The wisdom of that perspective will forever reside in the "selfless, immeasurable mind"; the humanity of it will forever be reflected in a socially connected freedom.

This revolution, with no beginning and no end, will be its own validation. It will find its true test when it explains all relationships only to discover that they are never constant. Humanity will, at long last, understand eternity. As Emerson might put it, man and Nature are experimenters and endless seekers with no past at their backs.

The new science allows us to declare that the time for gambling with Nature and what makes us social processing humans is over! Either we educate one another to welcome the power of our social intellect and its inclination to relate to, empower, and free all phenomena; or we continue to hide behind a repetitive history driven by static motives and superstitions, forces that justify the reckless waste of what is possible and make inevitable the tribal conflicts in the economics, politics, and science that follow.

14

All phenomena are interdependent processes.

Beyond Chance

New variance replaces chance with the freeing excitement and vision that comes with possibilities. Indeed, in studying the new science of possibilities and its architecture, we gain skills, strategies, knowledge, and perspectives to do all of the following:

- Learn, understand, and act upon the implications of the possibilities of new variance;
- Establish and practice a growing welcoming for changeability;
- Expose the hiding places of the independent, isolated ego—the places where ignorance, fear, and apparentness hide—and fill these spaces with relating, empowering, and freeing.

By embracing interdependent-processing forces, we recognize the intellectual potential to represent how and why interdependent-processing relationships are the legacy of a sociogenic Nature; thus, all phenomena evolve to be universal. The developmental character of the new science models processing why *and* how phenomena unify to form basic elevating, generative cycles for all human and cosmic activity. Comprehending interdependent processing, and representing it, is the work of the new scientist as a seeker of meaning with these principles in mind:

- Assumptions based upon probability are, for the most part, conceptual.
- The community-at-large will most readily accept conceptual explanations.
- The conceptual rarely transforms persons and processes; it almost always fixes them.
- Conceptual persons and processes are non-persons and non-processes.
- Non-persons do not think and thus do not process.
- Markets can be isolated and functionally autonomous.
- Sciences can be isolated and functionally autonomous.

Possibilities owe their generous disposition to Nature's social processing character. With the maturation of cosmic time and space, the chaotic and determined universes are fully interdependent. They are one in their changing places within the power of social process-centricity. Together, they form asymmetrical interdependent phenomena. Such phenomena are the greatest of all possibilities, for without them there would be no dynamics and thus no time, space, changeability, universes, language, intelligence.

Factual, geometric "proofs" and chaotic, determined processes establish no lasting phenomena and no lasting explanations. Such "proofs" and chaotic, determined universes have a highly compatible relationship in which the chaotic is not immediately predictable but is determined. Within interdependent changeability conditions, dominant chaotic phases are, however, predictable. Within these conditions, gravity is relative. Relative gravity may well be the proof for complexities.

All at once:

- Social process-centricity $\rightarrow$ relative gravity.
- Social process-centricity $\leftrightarrow$ relative gravity.

These relationships fuel changeability possibilities. Expansion and collapsing rates are due to unequal asymmetrical universes redefining their interdependent complexities. They are all social process-centric *in origin,* and this explains how and why all universal entities are empowered and free to evolve and contribute to maturing levels of changeability. We may thus say the following:

- Whether moving or not, interdependent, asymmetrical, social processing phenomena are uniquely changing the infinitely complex universes.
- The increasing expansion rate of the universe may be a phase whereby its outer reaches are approaching and simultaneously moving away from other universes all at once.

Clearly, we are not more than biographers recording the universal long-term phenomenal activities of Nature that are coming together to allow a glimpse of maturing, diverse, asymmetrical cycles of Nature's identity. However short-lived these cycles, however enduring the elevated states of maturity, these asymmetrical cycles are the sources of infinite complexity, infinite possibilities, and, most important, new variance. *New variance is Nature's ultimate creation.* Initially chaotic, new variance is the legacy of possibilities and is therefore determined *and* predictable; and possibilities are the legacy of new variance. Consequently:

- All manner of new variance is developmentally both chaotic and relative, depending upon the reinventions of relationships among and between phenomenal processes following "brief" moments of singularity.
- All phenomenal processes, including time, space, motion, gravity, and mass, are variable, interdependent expressions of socio-cosmogenic process-centricity. Each owes its identity (life) to unequal interdependent processing with the others. They were born together and exist and change together. *No one expression of socio-cosmogenic process-centricity really exists.* Together, their interdependent processing is the cosmic choir.

15

Only leftovers have a possibilities destiny.

Thank God for Leftovers!

This cosmic experiment, this grand adventure, takes place in God's classroom. Until now we have been rather poor, isolated, competitive learners. The journey to and through the new science of possibilities is changing all that. Its architecture has become the lesson plan for a splendid emancipation. Freedom from the oppressive burdens of the competitive ego and probabilities allows a glimpse of a vigorous truth: *The end of ego-driven probabilities dogma marks the first true opportunity to be introduced to changing reality.*

Unequal merging and repelling forces are born from Nature's great intolerance for symmetrical equality, and these forces propel changing reality. So what we consider only probable and what we consider impossible are, in Nature, possible. They are possible because, as Dante put it, "Nature is the Art of God." And this means they must be free!

What John Adams said of freedom in human affairs is applicable to the cosmic: "Freedom is providential. Freed is Natural."

Whether developmentally merging or developmentally repelling (or doing both), all participants in Nature are empowered and free by virtue of their unequal, interdependent-processing relationships with one another. Equality leaves no space for freedom. Only "possibilities equality," being defined as full participation in unequal, interdependent processing, does.

We will never know and understand all of God's secrets. However, we can, by knowing and understanding some of God's operating principles, obtain an overview of God's blueprint:

1. Possibilities engender possibilities;
2. There are no categories of ideas and images in Nature that cannot be improved upon by relating to, merging with, and processing interdependently with the unequal asymmetrical interdependence of phenomena; all the mergings and repellings—all the processing—mark conceptions of new life.

Thus the new scientist grows increasingly open to the possibility that there is one unifying definition for everything. That definition is available only if and when we view phenomena moment to moment from the changing margin. From the changeability margin we see, through Nature's eyes, God's grand "freedom experiment." Empowered and free, Nature is God's handiwork. In its empowered freedom, Nature engenders interdependent, unequal changes—changes that merge and changes that repel. Both kinds of changes are developmental possibilities: they *generate* possibilities. They require only unequal, interdependent processing. Interdependence (unequal and asymmetrical) is the prepotent condition *and* standard for all creation. It is also the condition and standard for the short-lived tolerance for singular symmetry.

Although there is, in Nature, a greater tolerance for unequal symmetry, the latter too dissolves into brief moments of singularity. Moments of singularity, when they expire, expel antimatter destined to collide with positive matter. The resulting catastrophe leaves behind the ingredients for everything, for all phenomena! Leftovers!

All phenomena, however brief and enduring, are a function of their composition of unequal, interdependent leftovers: Possibilities Minds. This is the one unifying definition. It has the greatest promise of possibilities. It is God's promise and Nature's undertaking. This promise inhabits Nature's infinitely complex universes and their countless expressions of evolving interdependency. This promise is perspective beyond the borders:

> *"If you free yourself from self, you will see the universes as I am able to see them and as I am able to relate to them from the margin beyond the borders. You will acquire the potential to merge with everything of substance and spirit, most especially the leftovers. You will discover the source of my mysteries. You will begin to merge with me. Together we will rejoice in the freedom of ageless creations.*
>
> *"You are endowed—yes, empowered—with the processing wherewithal to relate beyond self to escape the confines of 'somewhere–nowhere,' which is an isolated body of one. Join me on the margin beyond the borders. There, together, we will see and discover the meaning of cosmic potentialities. We will be so free, we will create* **forever** *possibilities.*

"What you have never understood but must grow to understand is that, once creation is set in motion, my freedom is your freedom and your freedom is my freedom.

"So having merged with me, we will witness the eternal cosmic births of leftovers and internalize the interdependent processing with which leftovers germinate everything that is new. You need only to welcome interdependent membership in the social cosmic family of leftovers as I was 'born' to do *eternally.*

"Your pursuit for a place at the center of 'self' will be replaced with the meaning and *the promise of a special* freedom place *of pure clarity on the margin beyond the borders. You will be among the most rare of leftovers! The Possibilities Mind.*

"The greatest virtue and the source for clear perspective is the abandonment of ego. With ego, there is no place for me."

To take considerable liberties with a Freudian postulate, "The claims of our [probability] civilization [and the cooperating tyranny of ego] make life too hard for the greater part of humanity and so further [the] aversion to reality *and* the origin of neurosis… [thus, probability] neurosis seems to be a human privilege." The alternative image is this: just as the black hole is the caldron of the intolerant soup that nourishes antimatter for encounters that leave behind the essential leftovers for new creations, so the Possibilities Mind is the caldron for the processing soup that generates all manner of potentialities for creation.

Nature is the eternal release of unequal, asymmetrical, interdependent process-centricity. Because it is social, Nature is a perpetual new variance-engendering energy. Nature is God's perpetually unsettled, socially unbalanced grand experiment.

The new scientist's observations are therefore harvested on the margin, beyond borders. It is the only vantage for understanding the irrevocable and unequal reciprocity among the manifestations of all the forces of creation.

God empowered Nature with a selfless, social intellect. Consequently, God is the pre-eminent Possibilities Mind!

A selfless intellect empowers all of us, however unequal, to be free. Great changes and the greatest challenges take place:

> *"Once I was on a slow and anxious journey to someplace, anyplace. Then I embarked on an exhausting journey to nowhere. Now I am on an eternally renewing journey to everywhere. At long last, I learned everywhere is on the margin beyond the borders. It is the latitude and longitude for possibilities. Everywhere is home for leftovers and the Possibilities Mind."*

Nature's selfless, social intellect provides the new scientist with the experience to witness phenomena everywhere and the privilege to merge with all of them. This is how a unified science, the new science, evolves into clearer and more profound yet simple insights. New understanding follows:

- Stability has neither place for interdependent processing nor capacity to surrender to a dynamic reality.
- Reality, social reality, is the manifestation of interdependent-processing activity available for inclusive perspective and for reinventions.
- Social reality generates its own standards for changing measures for changing conditions.
- Interdependent participants travel standard measures of distance and time only in the context of small universal reality. Beyond, participants travel infinite and mutating degrees of random and relative interdependent relating in varied universal realities.
- Evolutionary and revolutionary insights are ideational statements of interdependence. Interdependence, which is by necessity unequal, fundamentally defines infinite realities everywhere. Reaching beyond standard measures of time and space (X = Y), each insight is a possibilities reality. Most important, possibilities realities overtake conventional descriptions of relationships and offer up fresh significance.

16

Only transmutation is immortal.

The Possibilities Mind II

The new scientist's primary challenge is to generate elevated and operationalized responses to the question "Why?" Now, as Nature reveals more of its social processing character, it is increasingly clear that understanding why probabilities science has become a superstition, and therefore cannot engender freedom, is long overdue. *In response,* the operational architecture of the new science of possibilities developmentally and systematically refines and even reinvents with its changeability relationships any content, ideas, representations, models, and equations that are substantive.

Possibilities science is more than a new paradigm, more than a new science that alters what most people suppose about creativity, discovery, perspective, meaning, and intellect. It is a new exportable community language technologized to merge with and process any changing substantive phenomena. Above all, it is a language of asymmetrical interdependence welcoming universal changeability as its indispensable equal *and* unequal companion-partner. The new scientific language is itself generative, innovative, explanatory, and social process-centric—an *evolving* means for reinventing elevated levels of intellect. It enables the search for social meaning by imaging perspectives with themes of intentionality: unifying mutating, diverse, asymmetrical, unequal, interdependent, *unprecedented* constellations. Such imaging offers orbits of potentialities fulfilling their human and humane, as well as cosmic, destinies.

In response, this evolutionary universal language, as it reinvents intellect, offers impact and understanding and meaning approaching the pure clarity of wisdom. It nourishes hopes, visions, and the best of our ideations; it is able to relate powerfully enough to hold all of us together in the accepting embrace of Nature.

There is a wonderful indeterminacy about the generativity of possibilities, of new responses, as we pass from insight to insight, action to action, relationship to relationship. All this occurs with an incessant ego-free enthusiasm and a grateful confidence in the social process-centric experience. The entire experience has personal as well as inclusive meaning.

The Possibilities Mind is unencumbered by the requirement to choose between probabilities and possibilities paradigms. Thus Nature, in all its complexity and changeability, lives in the Possibilities Mind. In this way, Nature reaches out to the possibilities scientist to offer its rewards—opportunities to share in the countless asymmetrical interdependent potentialities of the universes. Indeed, the mind is the primary laboratory of the possibilities scientist. Similarly, Einstein's laboratory was his mind: this bountiful banquet of potentialities was "the reason manifest in Nature." He discovered a marvelous harmony in this bounty, and generated a new dynamic reality made up of mutating, unequal participants.

As possibilities scientists, we simply draw the merging, evolving architecture, employing the printed page to help us proof images of the past, present, and future while we attempt to developmentally and systematically refine ideas, models, perspectives, and representational images. We do all this while freeing our Possibilities Minds from the bonds of conditioned language. We move from the laboratory of our minds to a new changeability language paradigm, reinventing and elevating the interdependent processing of the great processors. The probabilities alternative has been and is to move from an established conditioned language paradigm to appendages of established language. The former advances historical intellect; the latter attempts to traditionalize thinking. Over time, traditionalized thinking, with all its tyrannical conditioning, becomes superstition.

For the possibilities scientist, there is no requirement to choose between "the lab in the mind" and "the mind in the lab." The new scientist is free to employ any combination of these processes, but trusts "the lab in the mind" more because it engenders far more possibilities while engaging his or her most powerful instrument: empathy. Essentially, the new scientist is an empathic social scientist, much like a great mathematician who *when free of fixed symbols* treats mathematics as a social science. Both are free of the conflicts that emerge when introducing possibilities thinking to a conditioned probabilities mind: free of the conflict about maintaining the false promise of conditioned stability.

The new science is a dynamic, interchangeable response to Nature's generous offerings. These offerings tell a story about each cosmic participant engaging in a love affair with the potentialities of all cosmic, mutating participants. Whether expanding or collapsing (or doing both), this social bonding—this love affair—is the heart of Nature's intentionality; it is also the heart of human ideation.

We are reminded here of Marcus Aurelius's words "Our life is what our thoughts make of it"; also of Emerson's statement "Thought is great and swift and free, the light of the world and the chief glory of man." To these we must add from William Hazlitt, "Great thoughts reduced to practice become great acts," an insight that serves as an excellent commentary on the relationship among relating, empowering, freeing, science, and applied technologies, and above all, the generative relationship among responding, merging, and intervening. Indeed, Hazlitt passes on to us the meaning of all the legacies of the Possibilities Mind as it exercises its freedom to merge with universes of possibilities.

Two main themes now come together: that the evolution of the new science is expressed in modular form; how and why these modular images free the intellect to relate and merge with any evolving phenomena by exercising asymmetrical interdependent processing.

The convergence of these themes, we are told by Nature, allows us, at long last, to escape blind concessions to conditioned convention and its conversion to superstition. So we offer a science of merging process-centricity as the life force of freedom. Like Nature itself, this life force cannot be crushed or trampled under foot by contemporaries or copyists who deny potentialities. The architecture of the new science is so generous it offers generations to come the option to be intentional and ideational applied scientists so they can free up their Possibilities Minds.

Nature's offerings confirm in our "lab in the mind" that the new science has a grasp of Nature's intentional, ideational possibilities intelligence so that the true human legacy may begin. History will be far beyond the conventional; it will be a veritable constellation of new dimensions to explore, relate to, and merge with. These will, Nature promises, be places where the empowering gifts of reinvention provided by leftovers will not be feared, denied, and trampled. These new places will establish standards where conditioned mentalities no longer defend ignoring the prolific union of the ideal and the practical:

places where mindless assaults upon emerging elevated visions of human possibilities are not even part of recall. Such new places will break the bonds of stupefying cycles of inflation and recession, of war and peace, challenging the dissociation of growing wealth in the face of deepening conditions of social injustice and global political policy based upon bombing or not bombing. Most important, these new places will merge the powers resulting from the diverse, free expressions of interdependent processing that shatter the cycles of "Dark Ages" and strengthen the all-too-brief moments of enlightenment. These places of the love affairs between and among processing human communities and other processing phenomena will disclose that there is every reason to pursue the endless mysteries of the Possibilities Mind as it labors to comprehend the possibilities universes. These places of union will constitute immaculate freedom.

For example, the numerous applications of possibilities science (education, economics, community justice, leadership, management, policy, and new capital development, to name but a few) and the detailed, operational architecture created for each in the process present collectively a possibilities paradigm for *a pure and free relating, thinking, and caring humanity. For the first time, Nature, humanity, and science emerge from the womb where achievable potentialities are germinated.* They, the potentialities, are provided with the meaning and the means to progress toward higher, elevated levels of actualization.

The leftovers (found and lost variance) are systematically defined to apply to maturing levels of the marriage between "pure" science and applied science. For the first time, all science is applied science! In this context, Einstein contributes still another lesson congruent with the social character of Nature, perhaps greater than $E = mc^2$: "The true value of a human being is determined by the measure and sense in which the individual has attained liberation from the self." Surely, then, applied genius is beyond traditional measures if for no other reason than it *freely* offers the means and the reasons to awaken the applied genius in all of us. With such awakenings of caring and with productive applied genius, the great challenges of life, community, science, and technology will never injure. They will heal and nourish the fabric that joins the human mind, spirit, and experience while providing the new variance to create opportunities for discovery, utility, unity, *and* meaning.

Humankind may yet become anything it is able to observe and merge with; moreover, it may be wise enough not to shape such phenomena, but to welcome what their secrets make of us. This is untainted freedom!

Still more from Nature: *"There is no cosmic ego, so if you are to maintain this good and great and generous relationship with me, the science you postulate must be humble."*

The new science is a humble science. It includes humanity among, not above, all other processing phenomena. As such, humanity joins in the multitude of unequal, asymmetrical, interdependent mergings. If humanity has any advantage over other creatures, we owe that advantage to the promise of a rapidly processing social character that has the potential to reflect and merge with the universal character of Nature. The new science is a universal social science. If we study its developmental, deductive models and merge with them, we merge with Nature.

In all its diversity and complexity, Nature insists: *"There is no wisdom or virtue in the pseudo-isolation brought about by a smothering fear, guilt, and guile promoting an oppressive individuality that assumes the fragile persona of superiority."*

As we study the new scientific models and understand the implication of their social changeability, it becomes increasingly vital to keep in mind that this science is not a singular strategy for dominance. On the contrary, it is an organic, multidimensional, social processing constitution. It is thus constituted to empower, free, and mature as it merges.

In a profound sense, this social, ideational, and intentional science has the will to render service, for it embodies the ethical. It is dedicated to servicing the intellect's discoveries, to expanding possibilities and the meaning of experience, and to servicing the maturing relationships within and between Nature and perspective.

17

The new science of possibilities will be continuously reborn in the Possibilities Mind.

Perspective Is Born in Relating

It has never taken very long for freedom's cry to give way to fear of change and thus surrender to an assumption—the threat of changeability—the opportunity to exercise freedom. This surrender is typically expressed in fractured, hysterical anxieties and neurotic denials of the fear of freedom. Sadly, this psychology and its aftermath of tribalism, sexism, and racism reside at the core of our less-than-human legacy. In short order, these diverse anxieties, fears, and neuroses come together to support oppression and waste the rich promise of leftover variance. *"Chances are,"* the best prediction of probabilities science, then reasserts itself as the standard cry, offering some illusory degree of safety from the adventures offered by change. *"Chances are"* becomes expressed vigorously when "results" are awash in failure, justifying failure.

In this context, the fear of change and the fear of freedom are the same neutering ailment, adding their negativism to the adoration of control, triviality, and artificiality. Perhaps this is how and why stability becomes the ideal commodity. Wherever and whenever these companions of the superficial and superstitious prevail, non-relating institutions and an isolating psychology are fabricated, defended, and marketed. Greater falsehood and pathology follow and become commonplace, while the reality of sound science, wisdom, and possibilities are subjected to vitriolic scorn. The process of "mind numbing" becomes the aspiration for major human-community enterprises, from child rearing to marketing to government. Tragically, science and technology are frequently called upon to supply the tools and method.

For the human community, Nature's incessant maturing and diversely complex changes may well be God's test to determine if we are worthy of meaningful universal experiences that enable the wisdom found in freedom and the perspective found in Nature.

Every new demanding change forms a new society, a new psychology, a new culture. New phenomena coalesce in functional, evolving truths. This is something to trust. The probability alternative is very sad indeed: "All things are likely to come to a barren end."

All phenomena relate and process and generate new beginnings because they are unequal, one to the other. From their inequality and disparity, they breathe life into pockets of possibilities cultures, human and cosmic. The new science of possibilities images *how*. The implications of the deductive changeability models provide *why*. *Possibilities science therefore promises nothing specific, but offers everything.*

By welcoming the lessons of Nature, the Possibilities Mind lends vitality to freedom's options. The greatest freedom is the uninhibited insights that proceed from a bolder, clearer, and more unifying view of the fluctuating but *developmental* interdependence between chaos and relativity, between relative gravity, light, energy, space, time. Like these forces of Nature, the Possibilities Mind is asymmetrical and, thus, not only free but empowered to be powerful. Explorations of the new scientific models are intimate snapshots of what goes on inside the empathic, universal Possibilities Mind. However brief the moments of these snapshots, they developmentally *nest* within each other, empowering and freeing the Possibilities Mind to be the midwife for the births of immeasurable potentialities. Ultimately, the Possibilities Mind frees the new science to be the midwife assisting in the births of infinite generations of Possibilities Minds.

We are thus freed from the repetitive cycles of tribal history based upon:

- An oppressed mentality governed by a conditioned, probabilistic, two-response repertoire;
- An exclusive focus on a finite resource masquerading as a value;
- The seduction of an incentive level of motivation that sustains isolated egos;
- An independence psychosis;
- Antisocial destruct systems that put a lid on human potentialities.

Nature imposes none of these limits. They are imposed and perpetuated by the probabilities flight from the freedom that comes from relating and processing.

Having engendered and housed the new science, the Possibilities Mind extends the power and thus the province of human intellect, human values, and human perspective, and while doing so, does the same for science. Now it is only a small effort for the Possibilities Minds to serve up possibilities philosophy. The legacy will, at long last, subvert the cycles of the rise and fall of cultures founded on a singular superstition about a finite resource, making those cultures minor, fading curiosities. Now the future, whatever its "becoming" manifestations may be, is worthy of a welcoming reception.

Humanity and its understanding of Nature are far less for having conceded to isolation by gambling on the small, fragile guesses of probabilities. The truth of probabilities is but partial and assumed; and as Alfred North Whitehead reminds us, "To know the truth partially is to distort the universe." Settling for partial truth inevitably culminates in superstition. Such truth is no better than a terrible psychosis and, like psychosis, becomes the primary instrument of the unthinking, non-relating spectator. It is no truth at all, and unworthy of possibilities.

Only relating to merge is worthy. Relating to merge is a concession to the universal social character and social intellect. Relating to merge with phenomena is the primal process enabling the empathic Possibilities Mind to become light, time, motion, gravity, energy, events, and potentialities of all descriptions, just as Einstein rode a beam of light.

18

Possibilities enhance Nature's potential and thus human potential.

Where There Is a Social Nature, There Are No Voids

For the human intellect and experience—indeed, for the human condition—the new science of possibilities provides vitality for perspective. Perspective, in all its growth-engendering forms, is as close to God as humanity can be:

- Perspective as meaning,
- Perspective as input,
- Perspective as process,
- Perspective as product,
- Perspective as feedback,
- Perspective as empowerment,
- Perspective as freedom,
- Perspective as evolving reality,
- Perspective as possibilities.

Such perspective is ongoing, contributing to interdependent process-centricity as the universal content.

The new science offers the simple clarity of an idea: *relating to merge*. This is the empathic first requirement for interdependent process-centric relationships, which, in turn, yield possibilities. Possibilities are, together, another simple but clear idea. They serve as embryos of possibilities ideas reflecting the Possibilities Mind's generativity.

Again, it all begins with relating. The terrible truth is, *isolation strangles everything*. Relating therefore recognizes truth and asserts it over partial truth, which, as we have seen, is no truth at all.

Each possibilities idea has the power to usher in a new history and so a new legacy for human potentialities. Each of Nature's possibilities exercises the power to usher in new legacies that alter the universes and their relationships with one another. *Possibilities thus alter the community of Nature*. In the absence of ego-dominated neuroses, each universe may be free of artificial repetition while maturing with achievable promise. To maintain and grow these degrees of freedom, the new scientist must come to understand what Arthur Schopenhauer understood: "To desire immortality is to desire the perpetuation of a great mistake." Everything in the past and everything now is only an approximation and therefore in error.

The new science may enable the courage to work, understand, and welcome *both* empowerment and freedom and hence overcome the defense of probability servitude—the wish "only for a just master," as Sallust put it. We must remember: "If thought is to become the possession of the many, and not the privilege of the few, we must have done with fear" (Bertrand Russell).

Whereas we will never know all God's secrets, we may have here an overview of God's *blueprint for Nature's work.* The new science heralds a knowable community of Nature: a community increasingly open and comfortable enough to celebrate the social character and requirements of Nature, to comprehend that consistency is, in Oscar Wilde's words, "simply a confession of failure"; a community in which the possibilities scientist continues to ask the question Whitehead raised, *"What is it all about?"*

This more knowable community of Nature, made up of diverse potentialities, will support learning, assisting its evolution and the spread of an organic, social psychology of interdependent processing that serves a free culture's free tradition. The possibilities culture, free and varied in provincial language and customs, will coalesce in function and process to produce the new wealth of social wisdom. Such wisdom, perpetually new, will be universes beyond the cultural mentality that prompted Wilde to write, "The public is wonderfully tolerant. They forgive everything except genius." Participants will come to internalize Wilde's principle "Nothing survives being thought of." Not even a possibility, we may add. This refinement, indeed reinvention of civilization, will be sustained in its growth by the cosmic value of the individual's immeasurable social intellect.

Free of "isms," this reinventing civilization will be precocious in its rapid facilitation and adaptation of advancing technology that translates into immediate awareness. At a primary level, this instant awareness of the miraculous range of potential for Nature's community allows generative alterations in the processing foundations of all institutions, basic assumptions, science, art, leisure, and life itself.

The most challenging of all the implications of these exploding awakenings are the ever more elevated requirements and conditions to be met for achieving intellectual, personal, and spiritual openness to the onrushing flood of possibilities ideas. The new science is again and again the possibilities midwife for these adjustments.

There is in this vision an ego-free innocence liberated from conditioned tribal guile, duplicity, chicanery, guilt, fear of freedom, and fear of relating to merge. There is in this vision only fine demonstrations of empowered freedom.

In the oppression of isolation, resistance to being free enough to relate, merge, and process interdependently is less than healthy. In this tomb of pathology, the fear of relating is equal to the fear of freedom. Freedom engenders relating. It is the eternal life-fulfilling cycle of Nature actualizing its promise. Here there is no place for the less-than-healthy to hide. In the eternal expanding and focused cycles of relating —*freedom*—there is nothing to hide and no place to hide. There are no dark corners!

Although "society often forgives the criminal, [but] never forgives the dreamer," as Wilde put it, the dreamer is the one who endures and who loves enough to stimulate genius to alter the community of Nature to be ever more comprehensible. And, we were there when Nature coached: *"Take encouragement from reminders that all the great processors were considered dreamers and vilified as leftovers. To this day, none has ever been forgiven for demonstrating what all of you can become."*

19

"The true God, the all-powerful God, is the God of ideas."

—Alfred Victor

Ideas From the Study of The Possibilities Mind

- Awareness that possibilities are the only justification for humanity;
- Awareness that possibilities are the enduring yet ever-evolving, multiplying core of universal development and its required vocabulary;
- Awareness that asymmetry is the basic texture of Nature;
- Awareness that "unequality" is the genius of everything;
- Awareness that equality is the full participation in generating possibilities. This is the only privilege!
- Awareness that processing is the caldron of the possibilities science, technologies, applications, and the transmutations of the Possibilities Mind;
- Awareness that higher orders of sociogenic nesting are the connective tissue of constantly new expressions of Nature;
- Awareness that discovery, insight, and perspective will come if the scientist is free of a demanding, asocial ego;
- Awareness that renewal emerges from merging and diverging reinventions of relationships;
- Awareness that possibilities science is a generous benefactor of probabilities science, providing a shelter from the absolute bondage of the asocial;
- Awareness that unequal reciprocity is the building block of the Possibilities Mind's anticipations;
- Awareness that an infinite multitude of relationships and a unified expression of these relationships are discovered in $X \neq Y$;
- Awareness that the most fundamental discrimination is between what is cosmologically social and absolutely asocial;

- Awareness that in unequal processing, inequality is the incalculable, powerful, and beautiful;
- Awareness that the exploration of chance demands a new consideration of relationships, and that the exploration of relationships demands a new consideration of chance—that each requires the other;
- Awareness that all creation is processing without a beginning and without an end;
- Awareness that the renewal of perspective grows out of its bond with interdependence;
- Awareness that fresh, new perspectives nourish the mind and the soul as well as *and even more than* all the teachings in all the universities;
- Awareness that there would be every reason to be neurotic if it were not for a processing science;
- Awareness that all that is required for enduring capacities to welcome change can be found in the view of the landscape of minorities from the margin beyond the borders;
- Awareness that what is most difficult about creating is the escape from the superfluous interruptions of a needful ego;
- Awareness that genius keeps reinventing reality just as Nature continues to reinvent reality;
- Awareness that each participant in the new science is nested, processed, and rotated in paradigms estimating a fragile, unifying arrangement;
- Awareness that the Possibilities Mind is itself a series of interdependent processes that merge for only an instant;
- Awareness that the Possibilities Mind provides rotating paradigms positioning possibilities for the reinvention of relationships;

- Awareness that the Possibilities Mind exercises its resolve, intensity, and powers best when the critics attempt to establish limits;
- Awareness that the poet, the artist, and the Possibilities Mind merge with and process with Nature;
- Awareness that at the moment of understanding phenomena, it is time to move on;
- Awareness that it is more difficult to follow phenomena than to fabricate phenomena, and that following is immeasurably more fruitful;
- Awareness that all the contributions of the Possibilities Mind mirror freedom as the highest value and the greatest all-conquering force;
- Awareness that the implication of all possibilities is the altering of perspective for the better;
- Awareness that the great history of science is replacing mind-numbing conditioned customs with new processing relationships;
- Awareness that most tradition and conditioned cultural customs are to be loved and left;
- Awareness that each possibility constructs a temporary universe—for better or worse!

What Claude Debussy said about music can be said of the love the Possibilities Mind has for the mind:

> *"I love music passionately. And because I love it, I try to free it from barren traditions that stifle it… It must never be shut in…"*

20

"The idea of God stands for the possible attempt at an impossible conception."

—Edgar Allan Poe

"We Do Not Arrive Here Alone"

We do not arrive here alone. We do not stay here alone. We cannot move on alone. We do not change alone. There are a multitude of companions before, during, and after a journey of discovery. They are a dynamic conglomerate, more than and less than one another. From moment to passing moment their interdependent-processing status mutates. This is the meaning and the legacy of the uncompromising social character of changeability. So it is with membership in the cosmos, and so it is with membership in human affairs. In the larger universes, and in the pathways between universes, membership in human affairs and membership in the cosmos are the same. *All participants will mutate, never to be the same; many will vanish.*

There is here, in these observations, something more fundamental than mathematics. Asymmetry, inequality, interdependent (unequal) processing, and the resulting unequal reciprocity establish the parenthood for change, for opportunity, and for possibilities. Interdependent processing, with its infinite power, generates infinitely increasing ideationally based phenomenal inventions and reinventions. We need only to be ego-free enough, social enough, and welcoming enough to get out of the way and stay out of the way. Nature counts on it!

Understood in the context of continuous interdependent processing, unequal reciprocity offers liberation to reach beyond the unreasonable and unreasoned adventurer. It is the law that determines relationship outcomes, and has been that law from the beginning, if there ever was a beginning. It is born and reborn in the inclusive and interdependent, ego-free and empathic processing of all possibilities. Unequal reciprocity is simultaneously the last *and* first phase of all creativity, and welcomed and respected as such. All phenomena, as they merge, re-emerge, diverge, are serviced by the inputs that shape and reshape the unequal reciprocity, sustaining possibilities reality.

The Possibilities Mind requires and is aware of companions: it processes, rotates, and nests the onrushing tide of the possible. In processing with each other and with the Possibilities Mind, the companions contribute to the explorations of what is developing and why. They endure with mutual gratitude.

These multidimensional relationships provide all of us with an evolving possibilities reality that becomes a shared possibilities sanity *and* intellect. It is all accomplished with what Nature teaches us about the relating, empowering, and freeing source of its energies: $X \neq Y$. We learn how Nature creates and what it creates. With this learning, the Possibilities Mind is empowered to boldly explore, understand, and act upon the unequal, asymmetrical possibilities of cosmic companionship.

Our private worlds and our shared universes are populated; probabilities perspectives and possibilities perspectives are populated. These near and distant companions are diversely multitalented, changing, and socially loyal. They all struggle with, suffer, and thrive differentially on the anxious adjustments that accompany shared, generative, unequal, reciprocal possibilities. Their unequal relationships mirror the who and what we and they may become.

In the process, these powerful but modest companions engender a possibilities social perspective on the margin beyond the borders, and thereby free the Possibilities Mind. The eventuality is that the Possibilities Mind fully internalizes new maturing obligations. They form the community mission to do the following:

- Promote the public nature of knowledge and expand the meaning of social processing perspectives;
- Trust with confidence that we will have yet more ideas and that they will be more powerful than the last;
- Pay the price to journey toward endless discoveries and follow the implications of endless possibilities;
- Respond and initiate to empower and free those beyond the boundaries of our own communities and tribes, and so nurture unique interdependent-processing environments;
- Fully employ an inclusive, faithful possibilities ethic and intellectual language as we progressively merge with all phenomena;
- Engage fully and ego-free in processing conversations about changing approximations of changing truths;

- Ultimately celebrate our shared universal social character by merging with, and thus contributing to, cosmic possibilities as we travel the connective bonding tissue between universes;
- Establish that our journey from universe to universe is the boundary-free reality of reasonable adventurers.

God employed Nature to blend unequal leftovers to create a wondrous variety of infinitely complex and socially connected universes. They stimulate the hunger to understand God. The great difference, then, between the probabilities mentality and the Possibilities Mind is the gap between little or no meaning and profoundly penetrating meaning.

For example, in terms of the traditional equation X = Y, the error variance, based on linear regression, is enormous! Even larger is the error variance in expressions of the relationship(s) among a number of variables (multiple R's). By comparison, in the context of X ≠ Y, the possibilities of the relationships among an infinite multitude of relationships *and* the possibilities meanings of a unified expression of *the* relationship become genuinely and generatively not only possible, but real. X ≠ Y is that unified expression.

Whereas the probabilities mentality is asocial and, for all large universal purposes, offers little or no meaning, the Possibilities Mind is cosmologically social. X ≠ Y represents the nuclear meaning and implications of unequal reciprocity with unlimited power as well as unlimited meaning.

Equality leaves only space for stasis—and a very small space indeed! So it is with absolute reciprocity. Equality (=), closed and limited in meaning, offers static and thus unnatural objectives: ultimate failure! Inequality (≠), open, social, and unlimited in meaning, offers ongoing processing. There is no failure!

Unequal reciprocity ensures growing degrees of freedom; that is, interdependent, unequal companions represent degrees of freedom. Degrees of freedom, in turn, represent interdependent, unequal companions. For the Possibilities Mind, each companion provides countless degrees of freedom. Probabilities science is limited to narrow (n-1) degrees of freedom, and hence a limited number of companions.

Such limited companions restrict implications, reduce predictive validity, and, most important, confine perspective. A confined perspective is the prepotent requirement for brain numbing, conditioning, social isolation, disempowerment, and control. Possibilities science provides increasing awareness of the growing numbers of interdependent, unequal companions, and therein resides the power of its perspective.

Once and for all, despite probabilities science's careful considerations of what is random, of chance, of the utility of the linear and curvilinear, and for all its classification of variables and standardized definitions of method, the science remains limited because it employs time-fixed data samples. The independent variables are confined, the intervening variables are "controlled" with rigid limits, and dependent variables are "free" only under controlled conditions! All these ingredients are measured with "scales" based upon predetermined suppositions: none are responsive to the universe—or, more accurately, *universes*—of changing, free possibilities. And the artificial control obstructs the generative, changing mind.

Again, even "degrees of freedom" are the products of fixed, "controlled," balanced equations based upon time-fixed artificial measures of central tendency. The "degrees of freedom" are never free; they represent no real phenomena; their artificial freedom can never directly aid in the processing of anything but artificial probabilities!

Probabilities suppositions are based upon the most elusive of all suppositions: what is "normal" and/or what is skewed in the context of two fixed dimensions—fixed in time! When probabilities science ventures into a third dimension (what is in reality multidimensional phenomena), it does so with "best fit" straight lines to connect with two other "best fit" straight lines, all at right angles one to another. *All this is an investment in establishing the validity of artificial independence!* Nothing is more artificially restrictive, controlling, and therefore less empowering than straight lines meeting to form right angles that do not exist anywhere! Equally unreal are curved lines reflecting time-fixed samples, controlled intervening variables, rigid independent variables, and the magnitude of *anything* but free dependent variables.

There is no freedom, no perspective, in contrived independence. There is only the foundation of every distortion. There is only isolation! Probability is, by itself, outside of the context of possibilities science; it is antisocial!

We have something much better than probabilities. We have a possibilities brain! With its hundred billion or more neurons and God-knows-how-many dendrites, the brain is empowered and free. It is, like the new science and all phenomena, a perpetual, social processing force. This force is the abundant dynamo servicing the Possibilities Mind of the new scientist.

If we wish to become new scientists, we must reach far beyond ego, beyond gender, beyond tribe, to merge with our unequal, social brains! This social brain, by its very nature, begets interdependent processing within itself and between itself and all processing phenomena. The new science operationalizes this interdependent processing. The architecture of the new science is itself, then, an interdependent changeability process: so is the brain and the Possibilities Mind.

Science is not any "thing." It is certainly not independent content. Nor is science merely the practice of specialized technologies, of semi-technical specialties. And it is surely not the mechanical manipulations of information technologies. Science is the reflection of infinite possibilities as it exercises its inclusive perpetual cycles of unequal processing. Perhaps the empowered and free brain, in its service to and interdependence with the new scientist's Possibilities Mind, is something better than science.

God wove the universes and the infinite cosmos from the evolving images exploding from the divine Possibilities Mind. The new science of possibilities is faithful to this ongoing genesis. In its mergings with all free phenomena, the new science is itself the legacy of possibilities. It is thus the first theory of True Science.

21

The innumerable connected universes are one for "us" and "them"; the connections are ours!

Toward Integrative Reinvention

Throughout history, integrative reinvention has been feared and denied and its promulgators viciously assaulted. This has been, and remains, a defensive reaction to integrative reinvention's power to expose, in its contributions and implications, the superstitions that drive the conventional. Superstition and convention, like all conditioned responses, prompt defense in the extreme, to the point of mindless cruelty. The time to break out of the controlling cycles of such defense, and to develop a new perspective, is long overdue.

In integrative reinvention, generative operational ideas follow generative operational ideas, with all engaged in interdependent (social) processing. Strung end to end, the processing moments create a developmental, cumulative perspective; this perspective is supported by increasing degrees of freedom provided by the growing awareness of interdependent, unequal companions. Each subsequent idea or insight is *nested* in a higher-order insight. Previously held notions are no longer the same: their social context has changed. They are now maturing outcomes of the developing social character. As such, they are immune to superstition (conditioned convention) and liberate processing intellects to reach freely beyond history's "successes" and "failures." When made public, integrative reinventions, in all their deductive, nesting, and rotating operational forms, are immediately utilitarian because they contribute to freeing the interdependent, unequal relationships within and between all phenomena.

Social processing's constantly emerging degrees of clarity originate from its congruent reflection of Nature's complex rhythms of continuous processing. This is the source of possibilities science's integrative reinventions of time, space, and even possibilities. Along the way, this emerging unity sweeps away the spinal-reflex conventions in the path of reason's power over mindless ideology, scientific and political.

Continuous processing is the source of the vitality and generative force of relating as it replaces numbing isolation. A repetitive conditioned history rooted in finite resources gives way to the maturity of renewed, reinvented infinite resources exploding from interdependent-processing relationships. Psychologically, the fear of freedom and possibilities is replaced by the appetite for the genius of a growing, inclusive perspective.

Probabilities alone have no tolerance for such genius. Probabilities may well be the source, foundation, and justification for conditioned convention when it is applied to understanding larger universes. In this sense, probabilities are betrayals of the social genius that is Nature. Betrayal in any form is a pathetic defender of the guilt that stupidity leaves behind. Betrayal only produces more stupidity.

Social processing is content. It is the content of social genius. For the human experience, it not only amplifies perspective but shelters reinvention and produces operational principles that obviate failure. Cycle after cycle, integrative reinvention offers progressive approximations of unity. The emerging unity is the function of perspective. That perspective is in no small measure dependent upon understanding two basic propositions:

- New variance is the product of unequal, interdependent-processing phenomena.
- Equality is full participation in interdependent processing.

Progressive expressions of unity produce diverse degrees of freedom and diverse degrees of focused activity: all relate to process the implications of diverse degrees of focused activity as a measure of time, space, and thus new variance.

Social processing is the universal intelligence. Without it, happenstance operates with mindless chaos. In the absence of social processing *and* unequal asymmetrical participants, there is only stasis. There is no place for intelligence. There is no place for volition. And it is unlikely that unity emerges from perpetual chaos in the shape of anything other than surrender.

Operational social processing as content drives integrative reinvention, and together they offer the birthplace for moments of discovery. Together, they are the human as well as universal legacy for all the "companions," pure and true. Herein is content with a soul. This is the fabric holding the past, present, and future of the universes on their unequal, changing, interdependent journeys. Only the privatized ego is an anathema to integrative reinvention. Beyond the privatized ego, we are able to welcome the merging resources of possibilities perspectives and shatter the cycles of the "Dark Ages," with their all-too-brief moments of enlightenment.

The empathic, relating science of possibilities is sustained by the same social processing that drives and operationalizes social genius, what becomes the behavioral expression of the social character of all companions and, thus, degrees of freedom.

We stand at the threshold of a new history: a history that, in the most basic sense, is freed by the new science. It will be civil, inclusive, generative, and, above all, fundamentally developmental, integrative, and social. The new science requires these criteria. The spirit of the reasonable adventurer requires these criteria. These criteria make processing knowledge and perspective public, and thus negate dynasties, classism, racism, and sexism and help to heal mentalities damaged by probabilities artificialities.

Once humanity discovers it is as free as other natural phenomena, there will be no reason, foolish or otherwise, to resist being one with the asymmetrical character of unequal, interdependent-processing relationships. All phenomena are sociogenetically disposed to relating, merging, and maturing with one another, whether the relating, merging, and maturing are periodic convergence or divergence. That is freedom! Nature is freedom!

The architecture of the new science presents the content for a humanity free of guile. It provides an intentional-ideational science beyond the confines of ego and conditioning and probabilities—a science that provides processing paradigms so that all science is ultimately applied science and, for the first time, science and humanity may merge.

Einstein declared, "It is in fact nothing short of a miracle that the modern methods of instruction have not yet entirely strangled the holy curiosity of inquiry." Sadly, too many of these methods are still anything but ego-free. Dominated by their conditioning to the probabilities built into the "normal" curve, they have no generative alternative but to put in place "chance" between discovery and intellect. In contrast, the new science freely offers the means and the meaning to awaken and free the applied genius in all of us. The great challenges of life and community will then never injure the fabric of curiosity, the spirit of relating and merging; rather, they will provide the opportunity and the reality for the possibilities of utility *and* unity.

The new science's models—its nested and rotating, interdependent, merging models for relating, empowering, and freeing—are together the transitional vehicles for our becoming anything we observe. We will be wiser for having done so if we do it not so much to shape phenomena as to welcome what their secrets might make of us if we were free.

If it is anything, the new science is a humble social science. It simply places humans among, not above, unequal natural and fabricated phenomena as unequal companions. It is in this changing context that the possibilities of the human intellect are most clear. It is in this context that the changeability learnings are and will forever be most profound. Again, if the human has any advantage at all over the beast, we owe that advantage to the promise of a processing sociogenetic character; that is, a character at one with the socio-cosmogenic processing character of Nature itself.

The Possibilities Mind, empowered and freed by the new science, is not a simple strategy to dominate: it is a humble approximation of the force of Nature. The Possibilities Mind is then an organic force that generates new options to empower and free the social, economic, and political, as well as the scientific, with the ongoing maturity of interdependent processing. So the intentional-ideational possibilities of the new science, of the Possibilities Mind, share the will, the means, and the meaning of rendering Nature's ever-unifying phenomena a service for discovery and expanding possibilities.

The province of intellect and values has thus been infinitely extended. The emerging perspectives will be increasingly more profound because of the inclusivity, diversity, unity, and growing purity of its welcoming wisdom. The future will be worthy of an eager, welcoming reception moment to challenging moment. And history, its old cycles shattered, will be a chronology recorded by inclusive wisdom indebted to diversity.

The passive-dependent unthinking and conditioned spectator-judge and the fact that we are a bit less for having conceded so fully and so long to the gambles of probabilities combine to explain the depression of inclusive wisdom; together, these culminate only in some fashionable superstition. Concession to our sociogenic character, to the social intellect that defines our evolving, rightful, unequal place in Nature, introduces us to the connective bonding pathways within and between universes. It is this social intellect, however diverse, that promotes experience full of the wonder of discovery; it reminds us of all the following:

- Independence and dependence are one and the same, with each simply employing a different language.
- Functioning at an interchangeable level alone, with content and/or anything else, makes no progressive contribution.
- Qualitative differences and insights depend upon relating and merging.
- Controlled, conditioned achievement is universally impossible.
- The free intellect, as in Nature, does not cooperate with controlled achievement.
- Only an empathic, ego-free intellect functions without neurotic qualifications.
- An incentive-based orientation is conditioned to sustain regressive relating.
- An incentive-based orientation is no more meaningful than random motivation; both fulfill the basic conditions for mindless, powerless, terminal conditioning.
- The free intellect, relating and merging, and Nature and processing are sustained in and between their unequal and symbiotic yet changing relationships, nesting and rotating.
- The ego is, above all, the most potent depressor variable.

There are so many forces—some bold, some subtle, and most seductive. The “reminders” above protect us against some of these. Here, the seductive forces are more than seducers: they are the jailers of reason. They may masquerade as bold or subtle as well as seductive. All jailers operate under the “protection” of what is conventional. All are intent on driving awareness away from our valid unequal experiences and toward the artificiality of equality. Each and together, they are fabrications. How many forces! How many translated into how many equations! How sad! How numbing!

Equality is such a tenacious apparition! Like some attractive but false messenger, equality promises to undo the validity of our experiential base by isolating the abundant mysteries of differences and distance. In its duplicity, equality strangely imposes the probable between intelligence and changing unifying truths. In truth, equality is a probabilities moment. It is a pilfering apparition, devoid of predictive and constructive validity and hence substance.

Because equality discredits meaning, it limits and distorts perspective. Perhaps most damaging, it undermines guile-free human contact and thus the social stuff from which love and genius are made. Most terrible of all, equality obliterates the fabric of substantive intimacy and its most precious gift: the reality of the immediate. All the while, and beyond, it exterminates absolutely the embracing, natural connectivity with all creation: the unequal and asymmetrical Possibilities Mind.

The Possibilities Mind processes the implications of inequality as follows:

- The modulating immediate is the only reality.
- Unequal diversity is the prepotent, renewing "resource force."
- Interdependence ultimately renders chance unlawful.
- Interdependence ultimately blends assumptions and implications.
- Every process continues toward convergence or divergence: convergence when developing toward singularity; divergence following the moment of singularity.
- The moment following singularity is occupied only with leftovers.
- Leftovers are the building processes for possibilities.

- There can only be changing statements of processing outcomes.
- Nature does not tolerate singularity because singularity holds the diversity of leftovers captive.
- Nature's social character employs inequality to generate new levels of diversity, which result in elevated levels of interdependence and the maturity of new responses and phenomena; all processing relationships yield new phenomena.
- Sociogenic Nature accounts for all changing realities: time, time/space relationships, and thus the dynamic fabric of CCBT, which provides cosmic unity.
- All positive, productive intentionality and insights are derived in the transition from the illusion of independence to operational interdependent processing.
- The unity of the ideal and the practical depends upon the sociogenic character of Nature.
- The variability of developmental merging is the governing principle for the creation of new variance and thus diversity.
- Evolving interdependent-processing relationships are both qualitative and quantitative and deliver a multitude of qualitative and quantitative possibilities.
- The intentional and ideational Nature of the new science validates the latter as a unified science sharing the same social character as Nature itself. The unity of the new science is imaged in the paradigms reflecting changeability's immeasurable contributions to empowering, freeing, and maturing phenomena. This imaging is clear, concrete, and operational, with representations of nesting and rotating functions, components, and processes.
- The greater the physical and/or psychological distance, the greater the apparent similarities shaping perspective, cosmic and human.

- The smaller the physical and/or psychological distance, the greater the differences shaping perspective, offering greater opportunities for differences to yield possibilities resources.
- The perspective provided by the new science—a perspective from the margin beyond the borders—presents great intimacy!
- Nature, with all of its sociogenic wisdom, moves to change all things.
- The shared fear of freedom and the subsequent focus on a finite resource account for why human history repeats itself.
- Absolute control requires the absurdity of absolute singularity and absolute symmetry.
- The diverse activity of inequality is the flexible glue holding the mutating universes to their infinitely complex possibilities relationships.
- Managing to empower, free, and mature, as does Nature, means elevated inclusive cycles of unequal, asymmetrical growth and unequal reciprocity.
- If successful, efforts toward symmetry yield absolute zero.

22

For all creation and its changing family of companions, "unequality" is more profound than cycles of chaos and relativity.

Freeing Possibilities

The Possibilities Mind asks:

> *"What is Nature?"*

Nature answers:

> *"I am elevated cycles of unequal, asymmetrical changeability possibilities."*

Nature asks:

> *"What is science?"*

The Possibilities Mind answers:

> *"Science is the study of the processing origins of possibilities generated by unequal, asymmetrical relationships within and especially between processing phenomena, and the implications of those relationships for evolving assumptions about yet-to-be universes of possibilities."*

When Newton employed mathematics *and* methodology to establish experimental measurement (the test of the validity of hypotheses), he endowed science with rigor. When the possibilities science merges interdependence and changeability to introduce new variance as the generative dynamo of Nature's social character, it reflects the endowment of all interdependent phenomena with unequal cycles of processing, nesting, and rotating. The new science represents infinite resources.

As we reflect upon the developmental images of the new science, it becomes ever clearer that the normal probabilities curve, based upon right-angle linearity, is artificial. When the curve is "represented" with more than two dimensions (variables), it becomes an increasingly debilitating distortion; there is no such thing as a single independent variable, let alone more than one. When the artificiality of probabilities paradigms is "institutionalized" as conventional thinking (as it has been), our relating is limited to predetermined and independent control functions—functions based on a false assumption.

Perhaps the larger implication here is that control requires dependency for the controller and the controlled, another bit of evidence reflecting the fear of freedom. By contrast, the Possibilities Mind internalizes the requirement of freeing to endow growth (merging and expanding) interdependently for all processing phenomena (functions, components, processes). *Control depends upon* $X = Y$ *equations, however artificial. Possibilities emerge from the freeing functions of* $X \neq Y$.

Sadly, as we have seen from our world's history and present condition, control as a product of fear or freedom (for others and self) produces something ugly. All political inventions to control "repetitive" cycles ultimately result in distortions of the interdependent social fabric of the community.

For human psychology, internalized inferiority is the parent of the pursuit of independence and the control that independence is assumed to make probable. Probabilities mentalities ignore or discount the value of the statistical exception: the leftovers. So preoccupied are they with artificial measures of central tendency and variance that even degrees of freedom are immediately attenuated. The basics of probability statistics are external expectations.

Relating awakens the generative power of unequal, interdependent processing. Relating is the fragile inception of freeing. Thus, the Possibilities Mind relates to, merges with, even pounces upon the statistical exception to understand its level of functions, empower it, and ultimately free it. Whatever the character of processing phenomena, phenomena will reconfigure as unequal leftovers with unequal leftovers. These reconfigurations provide renewing standards and conditions for possibilities functions, producing new variance diversity.

When Claude Monet wrote, "All I did was to look at what the universe showed me, and let my brush bear witness to it," he shared a wonderful and profound lesson for the would-be scientist-biographer of Nature. He was guided by the understanding that there is no ego-centric audacity in Nature, only new beginnings. This is how it must be for the scientist-biographer as well as the artist: both must be free of ego in an ego-free universe to create new universes.

Perhaps art or philosophy, far more than mathematics, is the pure applied science. Great works of visual and literary art, and great philosophical insights, are always an attempt to depict, in anything from plots to perspectives to propositions, the dynamic *chaos-relativity* partnership as Nature's substance and sustenance. In periodic and aperiodic rhythms, this partnership orchestrates the conditions and processes of freedom for all phenomena. Free phenomena support and nourish the complexity of the possible. There is no envelope!

In art, the demand for integrity is absolute, but the demand for truth is, wisely, only momentary. Integrity grows with an evolving, and therefore generative, truth. For traditional probabilities science, the demand for integrity and truth is too often politically confused with what is momentarily commercial. Given its divisions, the commercial is its only option while seeking some lasting "standstill" truth.

For better or worse, but mostly for worse, probabilities statistics have become something other than and more than a tool. Probabilities statistics have become a mentality in the service of controlling the power of diversity. This psychology has infested every major activity, including education, governance, commerce, economics, child rearing, science, and technology. If not for the burgeoning degrees of freedom that diversity provides for the creation of new phenomena, there would be no statistical exceptions. Neither would there be a Possibilities Mind.

At their integrated best, the human genius, spirit, and soul well up from the depth and growing breadth of their diversity. Such mergings of genius, spirit, and soul venture out with an inclusive ego-free empathy to generatively process the social character of what is changing. This diversity both demands and provides the abundant resources for curiosity as well as a welcoming, pure innocence. From its depth and breadth spring the new responses for constantly maturing freedom. It is at the social core of Nature's constitution. It is this source of cosmic freedom that energizes everything in God's universes. It is this freedom that energizes all of the many diverse expressions of genius, spirit, and soul. It lends its profound power to pathways upon which flows the evolution of the marvelous harmony of endless changing processes of a mutating reality. So the new science of possibilities is itself endlessly changing possibilities, sustaining possibilities perspectives and thus the Possibilities Mind.

We offer the Possibilities Mind as the caretaker of the generative power of diversity and subsequent new variance. The interdependent-processing relationship among and between diversity, new variance, and the Possibilities Mind is *the infinitely renewing cosmic resource.* Their interdependence in elevated cycles of processing not only render "isms" dwindling, fading curiosities but also fashion the shape and fabric of eternity's grand possibilities legacy.

Singularity freezes human genius, justice, and truth, as it does Nature. But diversity, with its new variance, sets them all free to be processed by the Possibilities Mind.

We thus postulate the following:

- Diversity/inequality drives the creation of infinitely new universes.
- Diversity/inequality transforms the functions and levels of functions of universes.
- Supported and empowered with n^{∞} degrees of freedom in the context of inequality, or $(X \neq Y)^{\infty}$, infinite universes and their changeability-identity images are a descriptive explanation of the infinitely complex universes.
- The awareness of infinite, interdependent-processing universes, of their changeability identities empowered with n^{∞} degrees of freedom in the context of cosmic inequality, constitutes the Possibilities Mind.

Our possibilities scientific platform is a changeability *nesting, rotating* unity. It begins and develops with our relating message from the past and our thinking-processing message from the future, and is fashioned with operational, intuitive paradigms.

This means that inequality interrupts not only singularity but also fixed cycles of chaos and fixed cycles of relativity. The initial chaos of unequal leftovers is the product of the inequality between matter and antimatter. The leftovers are free, however, to establish generative possibilities for infinite and unequal processing relationships. These relationships may evolve into elevated singularities that are interrupted by Nature's intolerance for singularity. They then disperse into countless leftovers.

Only in our small universe does $X = Y$ and $E = mc^2$. Equality is short-lived. Moments of equality are brief instances of probability masquerades.

$X \neq Y$ positions us where meaning is born and reborn: in the Possibilities Mind. It is where processing becomes the renewing dynamic. It is where culture as well as science, at the threshold of n^∞ degrees of freedom, are empowered and free.

Empowered with n^∞ degrees of freedom, culture and science—indeed *all* phenomena—engage in continuous interdependent and unequal processing. They become a continuously changing, self-organizing universe among companion universes. In this cosmic soup, $(X \neq Y)^\infty$ inequality allows subsequent unbalanced reciprocity to perpetuate inequality.

The Possibilities Mind instrumentally employs $(X \neq Y)^\infty$—inequality with infinite degrees of freedom—as the basis of a new mathematics of such inequality. The following definitions of phenomena introduce the primary functions and components of this new mathematics:

- All phenomena have processing functions.
- All phenomena have multidimensional components.
- All phenomena process interdependently.
- All phenomena exist in conditions of asymmetrical, curvilinear space. (They are self-organizing, interrelated phenomenal processing systems.)
- All phenomena may be measured by standards of changeability, with the following rules of changeability:
  - All phenomena are continuously processing.
  - All phenomena are unequal in the power of their processing.

- All phenomenal processing systems are nested in higher-order phenomenal processing systems.
- All lower-order phenomenal processing systems are genetically encoded by higher-order phenomenal processing systems.
- All phenomena rotate to become drivers between and among these processing systems.

23

In God's universe of universes, within and between relationships are $(X \neq Y)^{\infty}$ and $E \neq mc^2$.

The Masquerade Is Over

The intuitive Possibilities Mind flourishes when merging with changeability and reinvention. Within the operational paradigms of possibilities, small X = Y observations are empowered with increased degrees of freedom; that is, with $(X \neq Y)^\infty$, the Possibilities Mind becomes natural, and $(X \neq Y)^\infty$ is its inclusive yet focused platform. The $(X \neq Y)^\infty$ platform encompasses interdependence, processing, and changeability as well as nesting, encoding, rotating, and, of course, inequality.

- *The renewing insights of the Possibilities Mind, its $(X \neq Y)^\infty$ character, operate to relate to, merge with, empower, and free generative discriminations with infinite degrees of freedom.*
- *The relating, empowering, and freeing of unequal, interdependent-processing phenomena provides unequal reciprocity with* n^∞ *degrees of freedom for all cosmic companions!*

In traditional probabilities science, "findings" based upon n-1 degrees of freedom are artificial by definition and due to their derivation. Even $n!$ degrees of freedom (factorial designs) are artificial, having been based upon predetermined values of n. Thus, Nature's "no-failure inequality" operating principle moves beyond n-1 and $n^!$ degrees of freedom to n^∞ degrees of freedom.

Only n^∞ is congruent with the multitude of unequal processing cosmic companions. All these cosmic companions with infinitely changing functions of infinite levels demonstrate unequal reciprocity—the unequal reciprocity that establishes perpetual "unequality."

unequality ⟶ unequal reciprocity

Nature is forever unfinished!

However humble is unfinished, momentary humanity, God has allowed the human brain to develop into a hundred-billion-neuron wonder. With its possibilities, the Possibilities Mind may create infinitely complex universes anchored with infinite functions of infinite levels of changing, moving states. Considering this, humanity may not be one of God's more humble experiments!

The following levels of processing drive the Possibilities Mind, drive the experiment!

Level: Unequal Relating – $X \neq Y$

Level: Unequal Merged Relating – $(X \neq Y)!$

Level: Universally Inclusive, Free and Empowered, Unequal Merged Relating – $(X \neq Y)^{\infty}$

These interdependent and developmental levels of processing are accompanied by critical discriminations:

- That probability degrees of freedom are artificial: n-1
- That factored artificial probabilities degrees of freedom represent controlled freedom: (n-1)!
- That only infinite degrees of freedom reflect inclusively and perpetually free possibilities.

Although a human universe may be but a millisecond in God's infinite universe of universes, the selection of our universes matters. It is this selection that elevates and accelerates the evolution of civilization as well as scientific perspective.

For example, we may choose to dedicate ourselves to a particular processing function such as education. If we define that function in terms of learner processing functions, such as thinking skills, then we may develop a whole new human universe dedicated to education—a processing system whose dimensions (functions, components, processes, conditions, and standards) are themselves processing systems related interdependently.

Simply put, first we *nest* components such as curriculum information within our selected levels of functions; then we *nest* our instructional processes within our levels of components; next, we *nest* conditions such as relating, empowering, and freeing conditions within our levels of processes; finally, we *nest* our standards, such as diversity and changeability, within the conditions. We have now defined an entire universe:

- All of its functions are process-centric.
- All of its components are multidimensional.
- All of its processing is interdependent.
- All of its conditions are genetically encoded.
- All of its standards are rotated as drivers of processing.

This universe is as valid as any of God's universes. Indeed, for humankind's more immediate functions, it is more valuable than the very universal conditions in which it is nested. To be sure, in terms of humankind's long-term purposes, it will contribute profoundly to our appreciation of God's unfinished universes.

Inequality, or "unequality," within and between the universes of functions is the fundamental requirement for changeability and thus possibilities. Inequality within and between entities is misinterpreted as chaos. The distortion is an excuse for chaos, one in which even possibilities are mistakenly viewed as a reflection of chaos. Such distortion blocks the manifestations of possibilities and impedes our understanding of what possibilities do reflect: universes not simply expanding but exercising their unequal, interdependent social processing, as does the Possibilities Mind.

For example, representing any universe as a symmetrical sphere only contributes to establishing excuses for chaos devoid of its true meanings and the true meanings of the universes. Moreover, it imposes stasis upon changeability, which is better represented as a spiraling force, conveying movement and modification (see illustration on following page).

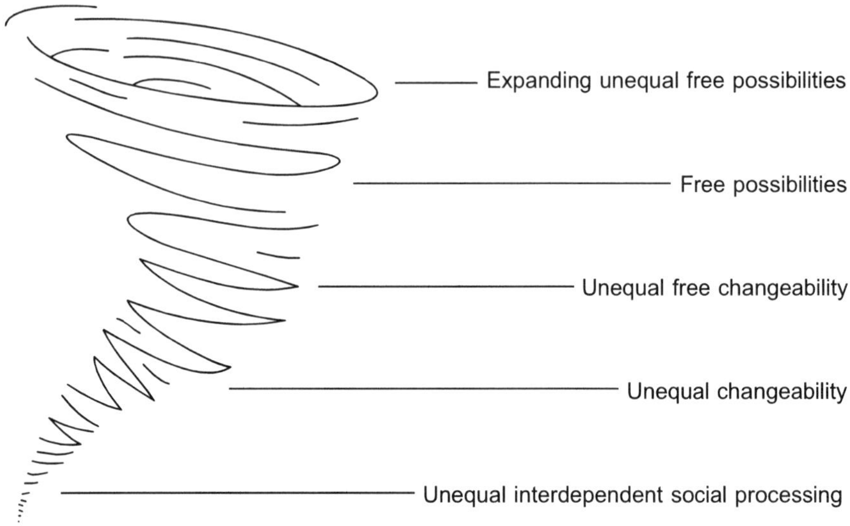

"Unequality" is the modifier.

These points are essential to our understanding possibilities universes and processing:

- Derived from the DNA of the Possibilities Mind, each arrangement of function *times* the number of components, *times* the number of processes, *times* the number of conditions, *times* the number of standards constitutes *and* perpetuates changeability (possibilities) universes.
- Together, the infinite arrangements approximate God's infinite universe of universes *and* their within and between changeability, processing relationships.
- These arrangements also define the changeability identity of universes uniquely and collectively.
- These arrangements provide the basis for predicting *and* accelerating the emergence of future identities and nesting arrangements.
- Any change occurring within a function, component, process, condition, and/or standard creates *a new universe.*

Most important:

- All functions, components, processes, conditions, and standards are social.
- The changeability arrangements of functions, components, processes, conditions, and standards provide the nesting, encoding, rotating DNA that evolving arrangements need in order to be the countless changeability, processing connections between universes.
- *The unequal functions, components, processes, conditions, and standards universes suggest chaos and relativity are misinterpretations of the dynamics of unequal nesting.*

All implications of implications owe their meaning to the following:

a. All nesting ingredients are social.
b. All nesting phenomena are social phenomena.
c. All nesting arrangements are unique universes.
d. Any transmutation within and/or between universes precipitates unique and collective changes throughout creation, human and cosmic.

Thus we account for the infinitely complex changeability identities of the infinitely complex universe of universes, human and cosmic.

Everything is made possible and *generative* with the power of $(X \neq Y)^{\infty}$ and the freedom afforded by n^{∞} degrees of freedom. This is the engine of the infinitely complex universe of universes. This is where the universes come from. This is where the Possibilities Mind lives. This is its essence.

- The unfinished conditions of the Possibilities Mind are the driving force *and* the consequence of its processing empowerment.
- The unfinished Possibilities Mind is made generative by the asymmetry of its processing. It is the natural paradigm.
- Unfinished conditions power changeability. So it is with the Possibilities Mind.

The Possibilities Mind accompanies exploding possibilities on their journeys. It can do nothing else. This is its function! This is the function of all universes. They are all unfinished. Even when universes implode and explode, they are in states of renewal and reinvention. So it is with the Possibilities Mind.

Renewal and reinvention fashion temporary truths and increasingly profound meaning. This is how all the cosmic companions from all parts of creation, from different universes, are connected.

For the Possibilities Mind, the connections are strands of diverse, renewing perspectives. They are the infinite, shared levels of functions of changeability. It is here where images of meaning emerge. With all their unique processing experience, the multitudinous functions arrive at the same moment at the mirror of multidimensional images of changing meaning.

Alone, functions are each less-than-fleeting humble experiments. Alone, they are destined never to validate anything but their social isolation. Together and asymmetrically nested, they engender meaning as the infinitely changeable functions powering creation.

The asymmetrical, unequal nesting of the functions, components, processes, conditions, and standards of all Nature reflects the asymmetrical Possibilities Mind. Thus the human universes may be as vast as the empowered and free human capability to employ possibilities paradigms to process the asymmetrical, unequal nesting of infinite functions and their components, processes, conditions, standards.

This asymmetrical, unequal nesting is the required mathematics. It is empowered with universal asymmetrical inequality to infinite degrees of freedom $(X \neq Y)^{\infty}$, free of equations! The mathematics of probabilities, with its imposed artificial measures, is at best miniscule in application and meaning.

Meaning, profound meaning, is revealed in the implications of implications. *Implications are functions.* Implications reveal their possibilities through and with the asymmetrical, unequal mathematics of the unequal nesting of changeability functions, components, processes, conditions, and standards.

From the margin beyond the borders, the unique and collective meaning of the infinitely complex implications of the infinitely complex universes of God's universe is revealed as functions in the nesting and rotation of their unfinished states. So it is with the Possibilities Mind.

As it is with Nature, the Possibilities Mind reveals its unfinished condition.

$(X \neq Y)^{\infty}$ is the power source.

24

There is no equality in all creation. Everything is unfinished and $(X \neq Y)^{\infty}$ is the first theory of science.

Momentary Functions Toward Meaning

For all the blessings of the unfinished, transmutating universes, we have a troublesome history. One of the most foreboding lessons of this history is that human probabilities mentalities transform into superstitions every idea, vision, technology, belief system, government, ideology, economic-commercial arrangement, and, hence, every educational model and culture.

There is an alternative. If we are willing to pay the price, we may choose the unfinished, inclusive, social processing dynamics of the Possibilities Mind. The price is realizing that the uncompromising essence of those dynamics thrives on the margin beyond the borders. This allows the new-science paradigms of the Possibilities Mind to respond interchangeably to Nature, to *be* interchangeable responses. In this way, Nature's interdependent social processing can be made public through the (asymmetrical) processing of the Possibilities Mind.

The Possibilities Mind not only bridges the gap between the human intellectual experience and Nature, but merges them. In this sense, in every sense, the Possibilities Mind and Nature are one.

Some of the most notable differences between our alternatives are shown in the illustrations on the following page. They make clear the contrasting consequences of our choice.

Science, culture, and civilization are at a crossroad, seemingly caught between the gambles of probabilities and the choices of changeability. We may continue to embrace conditioning, or we may embrace infinite interdependent-processing companions providing n^{∞} degrees of freedom. Here in universes where none of the multitude of processes is equal to any other, X = Y mathematics is fragile! Here the unequal companions all necessarily suffer and enjoy seemingly unfair but strangely just reciprocity. Their unequal changeability relating is more powerful, more profound than order. It is more powerful and profound than what is relative: cycles and phases of collapsing and expanding.

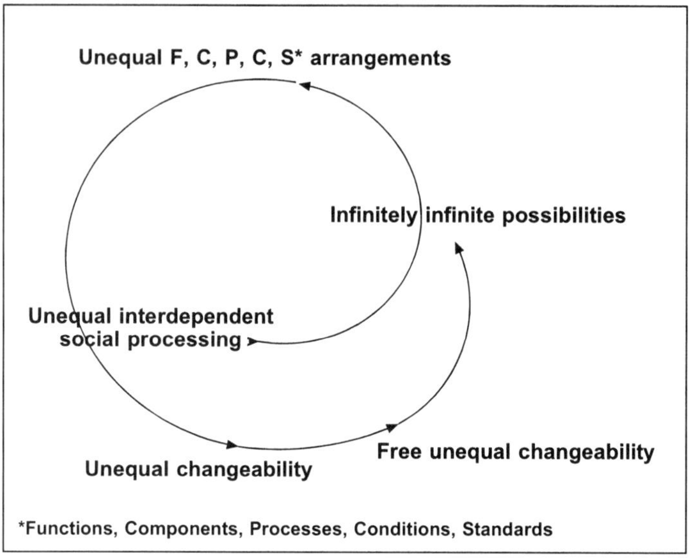

The Generative Social Character of Nature's Unequal Arrangements

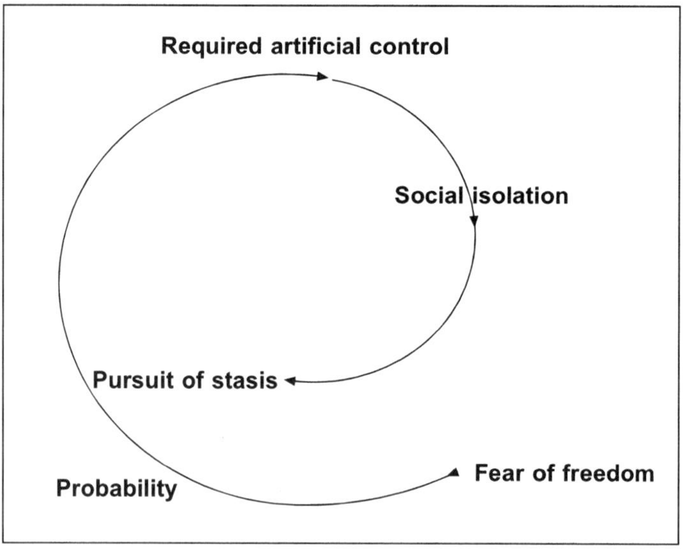

Human History Prior to the New Science of Possibilities

The inequality, the unfinished changeability, is Nature's seminal force generator. It *is* Nature's force. Multiplying possibilities, inequality is immune to control, resistant to partitioning. The common artificial variance that shelters confined probability is, by itself, alien. The unfinished, however, multiplies and nourishes the Possibilities Mind so intelligence and the universes are reinvented. While they, the universes, renew the evolution of companionships, they leave behind the static, superstitious, and artificial.

Merging with ever-larger unequal universes requires meaning pounded out of the processing functions of those universes. Now we may even anticipate yet-to-be perspectives. Whereas the brittle X = Y mathematics yields intolerable, unnatural dichotomies, unfinished changeability is to all creation "like water is to life." *All of us leftovers are worthy unequals!*

Some thousands of years ago, a processing problem-solver initiated the invention of written language. Now, the new science of possibilities continues that tradition as language continues to evolve. Such evolution sanctions *breakthrough thinking* in science, technology, social institutions, and, most important, unifying perspectives. These diverse and expanding yet temporarily unifying perspectives move from the rigid circular toward the visions of possibilities beyond the symmetrical confines of a repetitive history.

The language of the new science, and hence the Possibilities Mind, is social. It is operational beyond symmetry. It is dynamic beyond equations. Its perspective is vast, beyond a view from the center, beyond a view from the borders.

Representing universes as symmetrical spheres may lend itself to perspective, but it is wrong. There is no evidence that universes are symmetrical and spherical, or that they are expanding or collapsing symmetrically and spherically; quite the contrary! Also, the sphere is commonly accepted as the most universally primal phenomenon in terms of form. This too is wrong!

The assumption that universes are symmetrical and spherical defines an initial condition; thus, every subsequent "discovery" is encoded within the sphere. In this way, identifying universes as spheres becomes a conditioned "correct" response.

The borders of the universes, if indeed they have borders, are likely to be irregular, fluctuating and pulsating with evolving, changing possibilities. And the only primal phenomenon in the universes, in terms of anything, is change.

Unfinished universes may be anything changing: any form, weight, mass, volume, energy. They will be forever unfinished. When represented as a symmetrical sphere they are presented as finished. All empowered and free processing phenomena take on their own changing form as responses to and with their unequal interdependent social processing. They become self-organizing. Efforts to control them are analogous to imaging universes as spheres. If the universes are spheres, they are completed forms and we would have hopes to control them, a rather less-than-sane notion. We would end up with the same wheel we started with, proving that we had learned nothing.

If we begin with a condition, we end up with the same condition. If we begin with a function, the unequal nesting arrangements of functions, components, processes, conditions, and standards evolve into any and all possibilities relative to the initially observed function. Encoding everything within a condition has no empowering and freeing outcome for universes differing in changing form, weight, mass, volume, and energy.

In the context of cosmic time, the human experience is illustrated as the width of a single hair. In the company of perpetually free-moving possibilities, the human experience, even presence, may be more, *much* more. Indeed, the paradigm of the asymmetrical universe of universes—a universe of unequal processing universes—promises to be, in the decades to come, most profound and freeing for human experience, and a potent force for engendering all-embracing processing platforms, from the scientific to the communal to the personal.

Thomas Cahill, in his book *The Gifts of the Jews*, reminds us, *"Without justice, there is no God."* To this we humbly add, without unequal and unfinished processing universes, there is no God, only the void.

25

From the margin beyond the borders...

A Space to Mingle Intellectually

At a time when too many souls are suffering from an overdose of greedy independence, the Possibilities Mind is especially extraordinary. It is a phenomenon of unencumbered perspective and represents the heart of a civilized community's inclusive spirit. Rotating and nesting, the Possibilities Mind weaves an altering connectivity.

Empowered with the architecture of the new science, the Possibilities Mind provides freeing responses to:

- Social isolation—*the* most vexing and complex human and scientific conundrum;
- Unequal, interdependent social processing—*the* universally common source for the expanding evolution of possibilities, human and otherwise.

With the mathematics of inequality, the Possibilities Mind gives us the first real theory of science for understanding and applying the unfinished state of creation.

With the Possibilities Mind, we embark on a journey of discovery:

- From the artificiality of probabilities to a natural mutating reality;
- From static assumptions based upon linearity and equality to renewing expeditions based upon cosmic diversity;
- From the fabrications of parametrics to a view of the wondrous non-parametric pathways between non-parametric universes, all connected with a marvelous changeability bonding tissue.
- From the isolation of specialty "sciences" toward the growing unity of ideational science.

Along the way, through all of these reflections, we discover the justice and evolving truths in changing, inclusive perspectives "from the margin beyond the borders."

Perhaps most exciting are the discoveries to be found in the unequal companionships between order and disorder, and between expanding degrees of freedom.

We are reminded again and again of the following:

- However short-lived, singularity is, in its demise, the birthplace for all new variance.
- New variance is Nature's "ultimate creation."
- All the "universes" and all of us are statistical exceptions: *left-overs!* This is a shared connection that constitutes the stuff for infinitely emerging unities, complexities, and possibilities!

While everything is running out, even time, everything is running on to become something else—continuously.

Whereas variance services probabilities, the inequality of change is nutrition for Nature's possibilities intelligence. Moreover, the Possibilities Mind is not, like variance, something to be partitioned and controlled, but an empowering teacher-scientist.

"Most of all, we are teachers," we stated early on in this work. As teacher-scientists we did this work because the teacher-scientist is an empowerer. The Possibilities Mind is an empowerer when it is employed to teach. Teaching to empower is more than an applied technology; it is a personal, socially inclusive mission to be financially, culturally, intellectually, and spiritually generative; to demystify.

Along the way we repeatedly gained insights evolving from the higher-order processing of the nesting, encoding, and rotating of our new science. Rotating is a culminating freeing of degrees of freedom providing the countless conditions for the least of all processors.

Although love is about relating and merging, love is impotent if it does not deliver empowerment and freeing. It is not enough for great teachers to share what they love; they must align with substance and learner functions and create new universes.

Each class, each seminar, is an introduction to a new universe. These new places are alive with the unfinished quality of inequality, changeability, and the nesting, encoding, and rotating of higher-order possibilities processing. Thus, our science merges with our love's focused passion for the science's transforming substance, and together they evolve into insights. The science and love are perpetually in each other's fluctuating, unequal, freely rotating company. Together, they have the weight and power to weave visions of possibilities universes far more unfinished than their fragile individuality. Together, they move on to shape the dynamic fabric of human and cosmic bonds that are constructive and natural. These connecting bonds endure and grow in power with the sustaining contributions from the alliances between the infinite degrees of freedom and the inequality of the Possibilities Mind's new science and love. However explosive the ensuing insights, they are never complete.

Enabled by such freedom and such insights, the empowering teacher-scientist's Possibilities Mind is more than personal and interpersonal; it is philosophical, utilitarian, and, above all, asymmetrical.

Great teachers understand that concepts by themselves are feeble. Such empowerers provide operational, efficient paradigms representing the inequality of nesting, encoding, and rotating as fundamental, generative, and natural. The empowerer's Possibilities Mind is the embodiment of unequal nesting, encoding, and rotating of increasingly higher-order processing. Thus, the empowering teacher-scientist's mind possesses *unfinished* weight, measure, space, energy, mass, visibility, changeability, and possibilities.

At our best, we are empowering teachers. At our best, we are Possibilities Minds. In this context, perhaps our greatest exemplars are the Jews, who have had the courage to internalize God's changeability and "grow" their perceptions accordingly. When everyone around them relied unquestioningly on conditioned responses, they processed their changing worlds and God (see T. Cahill's *The Gift of the Jews*). In so doing, they grew out of "correct" choices toward *creative choices.* From their multidimensional perspectives, they discovered and created many things, including the helping professions that require empathic relating, merging, processing, and initiative possibilities. It is this ability—to accept changeability and to develop the multidimensional perspectives it requires—that we are aiming at.

When we are at our intellectual best, we realize the best answer in response to a major issue is always "all of the above." Each challenge is transformed into a nesting, encoding, rotating processing experience—an experience of freeing, of growth. The alternative is the inevitable transformation of the "correct" response into bleak superstition.

In the larger context of the non-parametric space of larger universes, X = Y (the "correct" response) fails. In and between, nothing is equal. In small spaces, small universes, parametrics require equalities that define symmetrical areas founded on the artificiality of "reliable," "stable" measures of central tendency and "acceptable" estimates of assumed deviations.

In so many ways, the multitude of unequal companions are the empowering teachers of the Possibilities Mind.

References

Berenson, B. G., and Carkhuff, R. R. *The New Psychology of Possibilities: The Future of Psychology.* Amherst, MA: HRD Press, in press, 2001.

Boorstin, D. J. *The Discoverers.* New York: Vintage Books, 1985.

Boorstin, D. J. *The Creators.* New York: Vintage Books, 1993.

Bronowski, J. *The Ascent of Man.* Boston/Toronto: Little, Brown and Company, 1973.

Cahill, T. *The Gifts of the Jews: How a Tribe of Desert Nomads Changed the Way Everyone Thinks and Feels.* Anchor Books, 1998.

Calder, N. *Einstein's Universe.* New York: Greenwich House. Distributed by Crown Publishers, Inc., 1979.

Carkhuff, R. R. *Human Possibilities: Human Capital in the 21st Century.* Amherst, MA: HRD Press, 2000.

Carkhuff, R. R., and Berenson, B. G. *The New Science of Possibilities: Vol. I. The Processing Science.* Amherst, MA: HRD Press, 2000.

Carkhuff, R. R., and Berenson, B. G. *The New Science of Possibilities: Vol. II. The Processing Technologies.* Amherst, MA: HRD Press, 2000.

Carkhuff, R. R., and Berenson, B. G. *The Possibilities Leader: The New Science of Possibilities Management.* Amherst, MA: HRD Press, 2000.

Carkhuff, R. R., and Berenson, B. G. *The Possibilities Organization: The New Science of Possibilities Management.* Amherst, MA: HRD Press, 2000.

Clark, R. W. *Einstein: The Life and Times.* New York: Avon Books, 1971.

Einstein, A. *Out of My Later Years.* Secaucus, NJ: The Citadel Press, 1956.

Frank, P. "Einstein's Philosophy of Science." *Review of Modern Physics,* 21, no. 3 (July 1949).

Morgan, G. (Ed.). *Monet: The Artist Speaks.* San Francisco: Collins Publishers, 1996.

Seldes, G. *The Great Thoughts.* New York: Ballantine Books, 1985.

Wells, H. G. *The Outline of History: Being a Plain History of Life and Mankind, Vol. I.* Edition revised by Raymond Postgage. New York: Doubleday, 1961.

Acknowledgments

We acknowledge with enthusiastic appreciation all those who empowered and freed us to relate and merge with possibilities universes.

First among them is Sigmund Freud, who created a system to share the implications of physical, emotional, and intellectual possibilities interdependency for our history, our present, our future, and our individual social experience. Equally, we express our appreciation for the gifts that Albert Einstein bequeathed, enabling all who followed to alter and evolve changing perspectives on the nature of experience and the cosmos. Together, Freud and Einstein boldly merged science and the social psychology of experience. Without these giant reasonable adventurers, all of history and the future for humanity and the cosmos would be insensible. They taught and demonstrated the infinite possibilities inherent in the power of "the lab in the mind."

We owe an exceptional debt of gratitude to James Drasgow, not only for his contributions to the substance of this work, but for the sustaining force of his vision throughout the years. His generous, unfettered perspective, rich with scholarship, wisdom, and clarity, has made all the difference.

There are those friends and colleagues who patiently and fully participated in our often audacious explorations: Debbie Decker Anderson, George Banks, Dave Berenson, Joel Berenson, Sally Berenson, B. R. Bugelski, John Cannon, Irv Conrad, Ted W. Friel, Darryl Howard, Julie Howard, Cindy Littlefield, Fred Mackler, Susan Mackler, Shirley McCune, Richard Muise, Richard Pierce, and Jeanette Tamagini.

There are those who contributed to transforming early manuscripts into readable text: Don Benoit, Dave Burleigh, John Cannon, Bob Carkhuff, Jr., Chris Carkhuff, and Mary George (who offered sensitive editing with uncommon understanding of the substance).

Jean Miller deserves an exceptional note of recognition for implementing our "rapid prototyping" method of writing: about one dozen versions were produced before final copy.

Finally, we owe a debt of everlasting love and gratitude to those who have been absolute in their commitment to enabling us to actualize our vision: David N. Aspy for his path-finding educational applications of possibilities principles; Simon Berenson for his unfailing support throughout a lifetime; Dave Berenson for his courage and intelligence in generating some of the first true educational technologies; Andrew H. Griffin, Jr., for his life-long commitment to continuous interdependent processing; John T. Kelly for acting as our mentor and guide on this possibilities journey.

Above all, we wish to dedicate our work to those who supported it: our wives, Bernice and Gloria, who related to our experience, empowered our potential, and released us to the freedom of our scientific pursuits.